MATHEMATICS

IN

ARCHITECTURE

PRENTICE-HALL INTERNATIONAL SERIES IN ARCHITECTURE

Mario Salvadori, editor

SALVADORI *Mathematics in Architecture*
SALVADORI AND HELLER *Structure in Architecture*
SALVADORI AND LEVY *Structural Design in Architecture*

PRENTICE-HALL, INC.
PRENTICE-HALL INTERNATIONAL, INC., UNITED KINGDOM AND EIRE
PRENTICE-HALL OF CANADA, LTD., CANADA

PRENTICE-HALL, INC, ENGLEWOOD CLIFFS, NEW JERSEY

MARIO SALVADORI

Chairman, Division of Architectural Technology
Columbia University

MATHEMATICS IN ARCHITECTURE

PRENTICE-HALL INTERNATIONAL, INC., *London*
PRENTICE-HALL OF AUSTRALIA, PTY. LTD., *Sydney*
PRENTICE-HALL OF CANADA, LTD., *Toronto*
PRENTICE-HALL OF INDIA PRIVATE LTD., *New Delhi*
PRENTICE-HALL OF JAPAN, INC., *Tokyo*

The endpaper illustration is a reproduction of *Numbers,* a gouache by Walter Fusi.

Current printing (last digit):
10 9 8 7 6 5 4 3 2

Library of Congress Catalog Card Number 68-9381

Printed in the United States of America

PREFACE

This book has been written for those
who are afraid of mathematics,
who would like to learn how to use it for the purposes of architecture,
and
who do not want to spend more than a few weeks to achieve this goal.

It covers the basic mathematics required to obtain "ballpark" answers to most of the questions arising in architectural technology.

Discovering by himself, or with the help of a teacher, that it is easy to obtain quantitative answers to the problems of his daily practice, the architectural student will exorcise his fear of numbers and get ready to deal with the physical reality of architecture. He will learn at the same time the language of technology and thus establish the possibility of better communications with his consultants.

The fulfilment of these needs is an essential condition of good architecture in today's technological era.

New York, N.Y. MARIO SALVADORI

ACKNOWLEDGMENTS

The author wishes to express his gratitude to Professor Charles W. Thurston of Columbia University, who used this material in one of his courses, read critically the manuscript, and checked the solutions of the problems. His suggestions were most valuable.

Miss Amelia W. Bergvelt typed the manuscript with her usual extraordinary skill and care.

CONTENTS

1

ARITHMETIC

2

ALGEBRA

3

TRIGONOMETRY

4

ANALYTIC GEOMETRY

5

DIFFERENTIAL CALCULUS

6

INTEGRAL CALCULUS

CONCLUSIONS

APPENDIX

INDEX

MATHEMATICAL SYMBOLS

$=$	equal to
$\equiv$	equal by definition to
$\simeq$	approximately equal to
$\neq$	unequal to
$<$	smaller than
$>$	greater than
$\ll$	much smaller than
$\gg$	much greater than
$\therefore$	therefore

1

ARITHMETIC

1.1 INTRODUCTION

The numerical answers to the problems of architecture are obtained, mostly, by the four operations of arithmetic. These we learn in elementary school. But while everybody knows that $2 \times 3 = 6$, few of us are capable of evaluating quickly and correctly an expression like:

$$\frac{2{,}141 \times 0.00128}{0.0742 \times 13{,}642}. \tag{1.1.1}$$

In order to obtain "practically correct" numerical answers in calculations like those of (1.1.1) one must realize that:

(a) An accuracy of more than 5 to 10% is seldom required or warranted in our work;

(b) The most common and most dangerous arithmetical errors occur in locating the decimal point;

(c) A slide rule is the simplest and fastest tool for obtaining such answers.

A brief explanation of the slide rule is given in Appendix 1. You *must* read it *before* starting the next section. But this chapter will show you how to avoid mistakes and obtain quick correct answers, irrespective of the tool used to reach them. The fact that incorrect answers may be obtained in an infinite variety of ways, while there is only *one* correct answer, does not mean that your chance of getting the right answer is nil.

1.2 THE OPERATIONS OF ARITHMETIC

In order to satisfy conditions (a) and (b) mentioned in Section 1.1, it is most convenient to write numbers as products of powers of 10 times a number between 1 and 10 and to round off this last number to 2 or at most 3 figures. In so doing keep in mind that:

$$\begin{aligned}
&\cdots\cdots\cdots\cdots\cdots \\
10^7 &= 10{,}000{,}000 \\
10^6 &= 1{,}000{,}000 \\
10^5 &= 100{,}000 \\
10^4 &= 10{,}000 \\
10^3 &= 1{,}000 \\
10^2 &= 100 \\
10^1 &= 10 \\
10^0 &\equiv 1 \\
10^{-1} &= 0.1 \\
10^{-2} &= 0.01 \\
10^{-3} &= 0.001 \\
10^{-4} &= 0.0001 \\
&\cdots\cdots\cdots\cdots\cdots
\end{aligned}$$

and remember that when the dot is shifted to the left by n places the number is to be multiplied by 10^n, and when the dot is shifted to the right by n places the number is to be divided by 10^n, that is, multiplied by 10^{-n} (see Section 2.3).

Thus, in expression (1.1.1) we may write:

$$2{,}141 \simeq 2{,}140 = 2.14 \times 1{,}000 = 2.14 \times 10^3;$$

$$0.001_{\wedge}28 = \frac{1.28}{1{,}000} = 1.28 \times 10^{-3};$$

$$0.07_{\wedge}42 = 7.42 \times 10^{-2}; \qquad 1_{\wedge}3{,}642 \simeq 1.36 \times 10^{4}.$$

The value of (1.1.1) is then, approximately:

$$\frac{2.14 \times 10^{3} \times \cancel{1.28} \times 10^{-3}}{7.42 \times 10^{-2} \times \cancel{1.36} \times 10^{4}}. \tag{1.2.1}$$

The exponents of the powers of 10 in the numerator are *added algebraically*; those in the denominator are *subtracted algebraically* (see Section 2.3):

$$[3 + (-3)] - [(-2) + 4] = [3 - 3] - [-2 + 4] = 0 + 2 - 4 = -2.*$$

The result may be estimated at sight:

$$\frac{2.14 \times \cancel{1.28}}{7.42 \times \cancel{1.36}} \times 10^{-2} \simeq \frac{2}{7} \times 10^{-2} = 0.28 \times 10^{-2} = 0.0028.$$

The actual slide-rule accuracy result is:

$$0.00271,$$

while the result obtained on a desk computer carrying all three digits in each number is:

$$0.0027144.$$

The estimated result is 3.2% off; the slide-rule result, 0.16% off.

Try your hand at the following expressions, and obtain the given correct answers by the procedure outlined above.

Note: Numbers ending in 5 must be rounded *up* if the last but one digit is *odd*, *down* if it is *even*: 13.5 = 14; 12.5 = 12.

1.2.1 $0.75 \times 1{,}200 \times 348 = 3.13 \times 10^{5} = 313{,}000.$

1.2.2 $0.021 \times 0.04 \times 32 = 26.9 \times 10^{-3} = 0.0269.$

1.2.3 $\dfrac{0.021}{7.22} = 0.291 \times 10^{-2} = 0.00291.$

*Remember the rule of signs:

$$(+) \times (+) = +; \quad (+) \times (-) = -; \quad (-) \times (-) = +.$$

For example:

$$(2) \times (3) = 6; \quad (2) \times (-3) = -6; \quad (-2) \times (-3) = +6.$$

1.2.4 $\dfrac{0.0044 \times 1{,}724{,}000}{3.14 \times 200} = 1.10 \times 10 = 11.0.$

1.2.5 $\dfrac{722 \times 345 \times 12.1}{4.12 \times 0.082} = 0.892 \times 10^7 = 8{,}920{,}000.$

Lest you should feel that this kind of evaluation is not to be found in architecture, solve the following problem.

1.2.6 Each floor of a rectangular concrete building 155 ft × 127 ft weighs 250 lb/sq ft (including the weight of the columns). The building has 32 stories. Compute the total weight of the building in tons (there are 2,000 lb in a ton).

If concrete weighs 150 lb/cu ft, how many cubic yards of concrete are needed to erect the building? (there are $3^3 = 27$ cu ft in a cu yd)

If the soil on which the building is to be built can carry 1 ton/sq ft = 2,000 lb/sq ft, what is the required surface of the foundation?

Using 2 figures we obtain:

$$\begin{aligned}
\text{Area of one floor} &= 155 \times 127 = 1.55 \times 10^2 \times 1.27 \times 10^2 \\
&\simeq 1.6 \times 1.3 \times 10^4 \\
&= 2.1 \times 10^4 \text{ sq ft.}
\end{aligned}$$

$$\begin{aligned}
\text{Area of 32 floors} &= 2.1 \times 10^4 \times 3.2 \times 10 \\
&= 6.7 \times 10^5 \text{ sq ft.}
\end{aligned}$$

$$\begin{aligned}
\text{Weight of 32 stories} &= 2.5 \times 10^2 \times 6.7 \times 10^5 \\
&= 17 \times 10^7 \text{ lb} \\
&= 8.5 \times 10^4 \text{ tons} \\
&= 85{,}000 \text{ tons.}
\end{aligned}$$

(In a single operation to 3 figures we obtain:

$$\begin{aligned}
\text{Weight of the building} &= 155 \times 127 \times 32 \times 250 \\
&= 1.55 \times 10^2 \times 1.27 \times 10^2 \times 3.2 \times 10 \times 2.5 \times 10^2 \\
&= 15.7 \times 10^7 \text{ lb} \\
&= 7.85 \times 10^4 \text{ tons} \\
&= 78{,}500 \text{ tons.)}
\end{aligned}$$

$$\begin{aligned}
\text{Volume of concrete} &= \frac{17 \times 10^7}{1.5 \times 10^2} \\
&= 11 \times 10^5 \text{ cu ft} \\
&= \frac{11 \times 10^5}{27} \text{ cu yd} \\
&= 0.41 \times 10^5 \\
&= 41{,}000 \text{ cu yd.}
\end{aligned}$$

$$\text{Area of foundation} = \frac{17 \times 10^7}{2 \times 10^3}$$
$$= 8.5 \times 10^4$$
$$= 85{,}000 \text{ sq ft.}$$

1.3 POWERS AND ROOTS*

The symbol 3^4 is a shorthand notation for $3 \times 3 \times 3 \times 3 = 81$. The product of two powers of the same number has an exponent equal to the sum of the exponents of the two powers:

$$3^2 \times 3^3 = (3 \times 3) \times (3 \times 3 \times 3) = 3 \times 3 \times 3 \times 3 \times 3 = 3^5 = 3^{2+3}.$$

Negative integer exponents are defined by noticing that, by the rule for the product of powers:

$$3^4 \times 3^{-3} = 3^{4+(-3)} = 3^{4-3} = 3^1 = 3,$$

so that dividing both sides by 3^4:

$$3^{-3} = \frac{3}{3^4} = \frac{1}{3^3}.$$

Thus:

$$2^4 \times 2^2 = 2^{4+2} = 2^6 = 64; \qquad 2^4 \times 2^{-2} = 2^{4-2} = 2^2 = 4;$$
$$2^2 \times 2^{-4} = 2^{2-4} = 2^{-2} = \frac{1}{2^2} = \frac{1}{4}.$$

The power of a power is obtained by multiplying the exponents:

$$(3^2)^2 = (9)^2 = 81 = 3^4 = 3^{2\times2};$$
$$(2^2)^3 = (4)^3 = 64 = 2^6 = 2^{2\times3}.$$

By the definition of root and the rule for powers of powers, one may define a root as a *fractional power* (see Section 2.3), and one may write:

$$\sqrt[4]{81} = (81)^{1/4} = 3,$$
$$\sqrt{9} = \sqrt[2]{9} = (9)^{1/2} = 3,$$
$$\sqrt[3]{8} = (8)^{1/3} = 2.$$

*See also Section 2.3.

The most commonly encountered roots are square roots, cube roots and fourth roots. A table of square roots and cube roots appears in Appendix 3 (Table 1) and must be used in connection with "powers-of-ten" numbers. For example:

$$\sqrt{30} = 5.48; \qquad \sqrt{3{,}000} = \sqrt{30 \times 10^2} = \sqrt{30} \times \sqrt{10^2}$$
$$= \sqrt{30} \times 10 = 54.8;$$
$$\sqrt{3} = 1.73; \qquad \sqrt{0.0003} = \sqrt{3 \times 10^{-4}} = \sqrt{3} \times \sqrt{10^{-4}}$$
$$= \sqrt{3} \times (10^{-4})^{1/2}$$
$$= 1.73 \times 10^{-2} = 0.0173;$$

$$\sqrt{0.003} = \sqrt{30 \times 10^{-4}} = \sqrt{30} \times \sqrt{10^{-4}} = 5.48 \times 10^{-2} = 0.0548;$$
$$\sqrt[3]{80} = 4.3; \qquad \sqrt[3]{0.08} = \sqrt[3]{80 \times 10^{-3}} = \sqrt[3]{80} \times 10^{-1} = 4.3 \times 10^{-1}$$
$$= 0.43;$$
$$\sqrt[3]{0.80} = \sqrt[3]{80 \times 10 \times 10^{-3}} = \sqrt[3]{80 \times 10} \times 10^{-1} = 9.3 \times 10^{-1} = 0.93;$$
$$\sqrt[3]{0.008} = \sqrt[3]{8 \times 10^{-3}} = \sqrt[3]{8} \times 10^{-1} = 2 \times 10^{-1} = 0.2.$$

1.3.1 A square area is 450 sq ft. Evaluate its side a:

$$a^2 = 450; \; a = \sqrt{450} = \sqrt{4.5 \times 10^2} = \sqrt{4.5} \times \sqrt{10^2} = \sqrt{4.5} \times (10^2)^{1/2}$$
$$= 2.12 \times 10 = 21.2 \text{ ft.}$$

1.3.2 A foundation block is a cube with a volume of 9,800 cu ft. Find its side a:

$$a = \sqrt[3]{9{,}800} = \sqrt[3]{9.8 \times 10^3} = 2.14 \times 10 = 21.4 \text{ ft.}$$

1.3.3 A block of granite has a volume of 375 cu ft and a height of 5 ft. Find the side a of its square base:

$$a^2 = \frac{375}{5} = 75; \qquad a = \sqrt{75} = 8.66 \text{ ft.}$$

Check the following problems by means of Table 1 in Appendix 3.

1.3.4 $\sqrt{14{,}000} = 118.$

1.3.5 $\sqrt[3]{0.007} = 0.191.$

1.3.6 $\sqrt[3]{760{,}000} = 913.$

1.3.7 $\sqrt{0.004} = 0.0632.$

1.3.8 $\sqrt{0.04} = 0.2.$

Note: Fourth roots are evaluated as square roots of square roots, since, for example,

$$\sqrt{\sqrt{81}} = (81^{1/2})^{1/2} = (81)^{1/2 \times 1/2} = 81^{1/4} = \sqrt[4]{81}.$$

$$\sqrt[4]{1{,}296} = \sqrt{\sqrt{1296}} = \sqrt{\sqrt{12.96 \times 100}} = \sqrt{\sqrt{12.96} \times \sqrt{100}}$$
$$= \sqrt{3.6 \times 10} = \sqrt{36} = 6.$$

1.3.9 $\sqrt[4]{0.0042} = 0.255.$

1.3.10 $\sqrt[4]{15} = 1.967$

1.3.11 $\sqrt[4]{17{,}000} = 11.42.$

1.3.12 Determine the diameter D of a circular steel rod of area $A = 0.54$ sq in.

$$A = \frac{\pi}{4} D^2 \qquad \therefore \quad D = \sqrt{\frac{4A}{\pi}} = \sqrt{\frac{4 \times 0.54}{3.14}} = 0.83 \text{ in.}$$

1.3.13 The volume V of a cube is 1.60 cu ft. Determine the area A of one of its sides.

$$V = a^3; \qquad a = \sqrt[3]{1.60} = \sqrt[3]{16 \times 10^{-1}} = \sqrt[3]{16 \times \frac{100}{10^3}} = \frac{1}{10}\sqrt[3]{16 \times 100}$$
$$= \frac{1}{10} \times 11.7 = 1.17 \text{ ft.}$$
$$A = a^2 = (1.17)^2 = 1.37 \text{ sq ft.}$$

1.3.14 A plot of land has the shape of a right triangle. The two perpendicular sides measure 105 ft and 230 ft. Determine the length L of the third side. By Pythagoras' theorem:

$$L^2 = 105^2 + 230^2 = (1.05 \times 10^2)^2 + (2.30 \times 10^2)^2$$
$$= 1.10 \times 10^4 + 5.29 \times 10^4 = 6.39 \times 10^4$$
$$= 63{,}900 \simeq 64{,}000 \text{ ft}^2,$$
$$L = \sqrt{64 \times 10 \times 10^2} = 10\sqrt{64 \times 10} = 10 \times 25.3 = 253 \text{ ft.}$$

1.3.15 The vertical struts of a suspension structure are stayed by cables attached 30 ft up the struts and 45 ft from their foot. Determine their length L. ($L = 54.1$ ft.)

2

ALGEBRA

2.1 INTRODUCTION

In arithmetic one performs operations on given numbers. The numbers may represent a physical or any other kind of quantity (like feet, pounds or dollars), but the operations are performed irrespective of the quantities represented. In other words, *numbers are abstractions which stand for real quantities.*

In algebra we use letters to symbolize numbers, we perform operations on the letters and give numerical values to the letters only at the end of the calculations. For example, if we must find the sum S of 3 times the square of a number n and then take the square root of S, we write:

$$S = n^2 + n^2 + n^2 = 3n^2;$$
$$\sqrt{S} = \sqrt{3n^2} = \sqrt{3} \times \sqrt{n^2} = \sqrt{3} \times n = 1.73n. \tag{2.1.1}$$

(Any other letter could have been used to indicate the number, so that (2.1.1) may also be written:

$$S = 3a^2; \qquad \sqrt{S} = 1.73a \quad \text{or} \quad S = 3z^2, \ \sqrt{S} = 1.73z.)$$

This result is true for any n and the evaluation of $\sqrt{S}$ is *faster* and *more accurate* by means of (2.1.1) than by the original definition of S. For example, for $n = 3$, the definition of S requires the following calculations:

$$n = 3; \ \ n^2 = 9; \ \ 3n^2 = 27; \ \ \sqrt{S} = \sqrt{27} = 5.19,$$

and, moreover, the evaluation of a different square root for each value of n, while (2.1.1) gives directly:

$$\sqrt{S} = 1.73 \times 3 = 5.19.$$

In algebra we go one step further in abstraction than in arithmetic, since the symbol n is an abstraction standing for *any* number, i.e., is an abstraction of an abstraction.

Once the abstract meaning of the letters used in algebra is understood, there is no difficulty in handling letters as if they were numbers. Just remember that letters eventually will become numbers and you will appreciate the efficiency of algebra.

2.2 THE SHORTHAND OF ALGEBRA

The basic shorthand of algebra consists in:

(a) dropping the multiplication sign between letters, and
(b) using brackets to indicate repetitive operations.

For example, the product "a times b times c" is written simply:

$$abc \equiv a \times b \times c. \tag{2.2.1}$$

This simplification, obviously, cannot be used with numbers since 3×2 is six and 32 is thirty-two. The sum of the products of a, b and c by the same number d is written by "factoring out" d, as:

$$da + db + dc \equiv d(a + b + c). \tag{2.2.2}$$

Brackets can be "nested" one inside another, but the operations they indicate must always be performed *starting from the inner bracket and working to the outer brackets:*

$$d[a(b + c)] = d[ab + ac] = dab + dac,$$
$$a[b + c(d + e)] = a[b + cd + ce] = ab + acd + ace.$$

Similarly, brackets may also be used to "factor out" a factor consisting of more than one letter which multiplies another factor of more than one letter:

$$ac + ad + bc + bd = a(c + d) + b(c + d) = (c + d)(a + b).$$

Thus, parentheses or brackets may be used to factor a "written-out" expression and a factored expression may be "written out":

$$a^2 + ac - ad = a \times a + a \times c - a \times d = a(a + c - d),$$
$$a[b + c(d + e)] = a[b + cd + ce] = ab + acd + ace.$$

Factoring or writing-out may lead to simplified expressions which, being valid for any values of the letters involved, are called *identities.* For example:

$$(a + b)(a - b) = a \times a + a \times (-b) + b \times a + b \times (-b)$$
$$= a^2 - ab + ba - b^2 = a^2 - b^2.$$

The useful identity:

$$a^2 - b^2 = (a + b)(a - b) \qquad (2.2.3)$$

leads to simplifications in expressions like:

$$\frac{(a^2 - b^2)}{a + b} = \frac{\cancel{(a + b)}(a - b)}{\cancel{(a + b)}} = a - b.$$

No simplification occurs in algebraic division unless common factors appear at numerator and denominator; thus:

$$\frac{a + b}{c + d}$$

cannot be simplified, but:

$$\frac{a^2 + ab}{ac + ad} = \frac{\cancel{a}(a + b)}{\cancel{a}(c + d)} = \frac{a + b}{c + d}$$

2.3 POWERS AND ROOTS

The product of n factors all equal to the same number a is abbreviated by the symbol:

$$a^n,$$

(devised by Descartes to abbreviate $a \times a \times a \times \cdots \times a$, n times), which, with m and n two positive integers, leads to:

$$a^m \times a^n = a^{m+n}, \tag{2.3.1}$$

$$(a^m)^n = a^{mn}. \tag{2.3.2}$$

For example:

$$a^3 \times a^2 = (a \times a \times a) \times (a \times a) = a^5 = a^{3+2};$$

$$(a^3)^2 = (a \times a \times a) \times (a \times a \times a) = a^6 = a^{3\times 2}.$$

By (2.3.2) the nth root of a number may be expressed as a *fractional power*. Let:

$$a^x \equiv \sqrt[n]{a}\,, \tag{a}$$

where x is unknown; then taking the nth power of both sides of (a):

$$a^{xn} = (\sqrt[n]{a}\,)^n = a = a^1.$$

Equating the exponents of a left and right:

$$xn = 1, \qquad x = \frac{1}{n}.$$

Hence, by (a):

$$\sqrt[n]{a} = a^{1/n}. \tag{2.3.3}$$

Similarly, by (2.3.1), negative exponents may be defined as follows. Let:

$$\frac{1}{a^n} \equiv a^x, \tag{b}$$

where x is unknown, and multiply both sides by a^n:

$$a^n \times \frac{1}{a^n} = a^n \times a^x \qquad \therefore \quad \frac{a^n}{a^n} = 1 = a^{n+x},$$

by (2.3.1).

But, by definition (provided $a \neq 0$):

$$a^0 \equiv 1.$$

Hence:

$$a^0 = a^{n+x} \qquad \therefore \quad 0 = n + x, \quad x = -n,$$

and by (b):

$$\frac{1}{a^n} = a^{-n}. \tag{2.3.4}$$

Combining (2.3.3) and (2.3.4):

$$a^{-1/n} = \frac{1}{a^{1/n}} = \frac{1}{\sqrt[n]{a}}. \tag{2.3.5}$$

Try your hand at the following expressions.

2.3.1 $\frac{a^r a^s}{a^r a^{-s}} = a^{2s}$.

2.3.2 $\frac{(a^r)^s}{a^r a^s} = a^{rs-r-s}$.

2.3.3 $\frac{(a^{1/r})^s}{(a^{1/s})^r} = a^{s/r-r/s}$.

2.3.4 $\frac{a^r + b^r}{(a+b)^r} = (a^r + b^r)(a+b)^{-r}$.

2.3.5 $\frac{a^{r-s}}{a^{r+s}} = \frac{1}{a^{2s}}$.

In this section we have not really added to our mathematical knowledge, but just introduced new shorthand symbols for powers and roots. On the other hand, these symbols give meaning to powers which were meaningless before. For example, a^2 means $a \times a$, but $a^{0.2}$ cannot be given meaning as a repeated product. Instead, since $0.2 = \frac{1}{5}$:

$$a^{0.2} = a^{1/5} = \sqrt[5]{a}.$$

Similarly:

$$a^{2/3} = (a^2)^{1/3} = \sqrt[3]{a^2},$$
$$a^{0.8} = a^{4/5} = \sqrt[5]{a^4}.$$

Evaluate the following powers.

2.3.6 $(9)^{-0.5} = \frac{1}{3} \simeq 0.333$.

2.3.7 $(9)^{1.5} = 27.$

2.3.8 $(8)^{0.67} \simeq 4.$

2.3.9 $(27)^{-1.33} \simeq \frac{1}{81} \simeq 0.0123.$

2.3.10 $(25)^{2.5} = 3125.$

2.3.11 $(81)^{-0.75} = 0.037.$

2.3.12 $a^{7.5} = a^7\sqrt{a}.$

2.3.13 $a^{6.5} = \dfrac{a^7}{\sqrt{a}}.$

The evaluation of a^m, where m is any decimal number, requires a table of logs. Problems requiring such evaluations occur seldom in elementary architectural technology, but a log table appears in Appendix 3 (Table 3).

2.4 POWERS OF $(a + b)$ AND THEIR APPROXIMATIONS

An important series of identities is derived by "writing out" the powers of the *binomial* $(a + b)$. For example:

$$(a + b) = (a + b)^1 = 1 \times a + 1 \times b,$$

$$(a + b)^2 = (a + b)(a + b) = a \times a + a \times b + b \times a + b \times b = a^2 + 2ab + b^2, \tag{2.4.1}$$

$$\begin{aligned}(a + b)^3 &= (a + b)(a + b)^2 = (a + b)(a^2 + 2ab + b^2) \\ &= a^3 + a \times 2ab + a \times b^2 + b \times a^2 + b \times 2ab + b^3 \\ &= a^3 + 3a^2b + 3ab^2 + b^3,\end{aligned} \tag{2.4.2}$$

$$(a + b)^4 = (a + b)(a + b)^3 = a^4 + 4a^3b + 6a^2b^2 + 4ab^3 + b^4. \tag{2.4.3}$$

You will observe that there is a pattern in these identities:

(a) Each term consists of the product of a power of a by a power of b.

(b) Starting with the highest power of a multiplied by 1, that is, by b^0, the successive terms are obtained by lowering the powers of a by 1 and increasing those of b by 1 so that the sum of the exponents always adds up to the exponent of the binomial.

$$
\begin{aligned}
(a-b)^2 = (a-b)(a-b) &= a^2 + a(-b) - ba + (-b)(-b) \\
&= a^2 - 2ab + b^2,
\end{aligned}
\tag{2.4.5}
$$

$$
\begin{aligned}
(a-b)^3 = (a-b)(a-b)^2 &= a(a^2 - 2ab + b^2) \\
- b(a^2 - 2ab + b^2) &= a^3 - 2a^2b + ab^2 \\
- ba^2 + 2ab^2 - b^3 &= a^3 - 3a^2b + 3ab^2 - b^3,
\end{aligned}
\tag{2.4.6}
$$

and have coefficients equal to those of $(a + b)^n$ but *alternating in sign.*

The corresponding Newton's binomial expansion becomes:

$$
\begin{aligned}
(a-b)^n = a^n - \frac{n}{1}a^{n-1}b + \frac{n(n-1)}{1 \times 2}a^{n-2}b^2 - \dots \\
+ (-1)^{n-1}\frac{n}{1}ab^{n-1} + (-1)^n b^n,
\end{aligned}
\tag{2.4.4a}
$$

where the powers of -1 in the last terms are used to obtain the correct signs of the coefficients.

Newton's binomial expansions lead to *very useful approximate formulas.* Let us notice that if b is much smaller than a the successive terms of $(a \pm b)^n$ decrease very rapidly in magnitude. For example, for $a = 100$, $b = 1$ we obtain from (2.4.1), (2.4.2), (2.4.3):

$$
\begin{aligned}
(100 + 1)^2 &= 100^2 + 2 \times 100 \times 1 + 1^2 \\
&= 10{,}000 + 200 + 1 = 10{,}201, \\
(100 + 1)^3 &= 100^3 + 3 \times 100^2 \times 1 + 3 \times 100 \times 1^2 + 1^3 \\
&= 1{,}000{,}000 + 30{,}000 + 300 + 1 = 1{,}030{,}301.
\end{aligned}
$$

It is seen that, within the accuracy required by most practical problems, the first two terms of the expansion give an acceptable result as soon as b is much smaller than a. Hence:

$$
(a + b)^n \simeq a^n + na^{n-1}b = a^n + n\frac{a^n}{a}b = a^n\left(1 + n\frac{b}{a}\right) \quad (b \ll a) \tag{2.4.7}
$$

For example, with the correct values to 4 decimals given in parentheses:

$$
(1.02)^3 = (1 + 0.02)^3 \simeq 1^3\left(1 + 3\frac{0.02}{1}\right) = 1.06 \quad (1.0612).
$$

$$
\begin{aligned}
(2.96)^4 = (3 - 0.04)^4 &= 3^4\left(1 + 4\frac{-0.04}{3}\right) \\
&= 81\left(1 - \frac{0.16}{3}\right) = 81 \times 0.947 = 76.71 \quad (76.7656).
\end{aligned}
$$

(c) The numerical coefficients of the successive terms are obtained by following triangle (called Tartaglia's or Pascal's triangle), in which e line starts with a 1 and contains the sum of the two adjoining numl of the preceding line.

BINOMIAL TRIANGLE

	Exponent
1	0
1 1	1
1 2 1	2
1 3 3 1	3
1 4 6 4 1	4
1 5 10 10 5 1	5
1 6 15 20 15 6 1	6

For example, $(a + b)^6$ has the terms:

$$a^6b^0,\ a^{6-1}b^1,\ a^{6-2}b^2,\ a^{6-3}b^3,\ a^{6-4}b^4,\ a^{6-5}b^5,\ a^{6-6}b^6$$

with the coefficients 1, 6, 15, 20, 15, 6, 1:

$$(a + b)^6 = a^6 + 6a^5b + 15a^4b^2 + 20a^3b^3 + 15a^2b^4 + 6ab^5 + b^6.$$

The coefficients of $(a + b)^n$ may also be obtained by noticing that the first is always 1, the second always $n = n/1$, the third $[n(n - 1)/1 \times 2]$, the fourth $[n(n - 1)(n - 2)/1 \times 2 \times 3]$, etc., down to 1 for the last. Thus Newton wrote $(a + b)^n$ in the following form, known as *Newton's binomial expansion:*

$$(a + b)^n = a^n + \frac{n}{1}a^{n-1}b + \frac{n(n-1)}{1 \times 2}a^{n-2}b^2 + \frac{n(n-1)(n-2)}{1 \times 2 \times 3}a^{n-3}b^3 + \dots + \frac{n}{1}ab^{n-1} + b^n. \tag{2.4.4}$$

For example:

$$(a + b)^4 = a^4 + \frac{4}{1}a^3b + \frac{4 \times 3}{1 \times 2}a^2b^2 + \frac{4 \times 3 \times 2}{1 \times 2 \times 3}ab^3 + \frac{4 \times 3 \times 2 \times 1}{1 \times 2 \times 3 \times 4}b^4$$
$$= a^4 + 4a^3b + 6a^2b^2 + 4ab^3 + b^4,$$

which checks (2.4.3).

If the binomial is $(a - b)$ rather than $(a + b)$, its successive powers are:

When n is *not a positive integer*, Newton's expansions (2.4.4) and (2.4.4a) never stop, since subtracting ones from n does not lead to a zero, but the expansions may be still shown to be valid when b is smaller than a. When n is not a positive integer, (2.4.7) leads to even more interesting approximate formulas for $b \ll a$. For example, for $n = -1$, (2.4.7) gives:

$$(a+b)^{-1} = a^{-1}\left(1 - \frac{b}{a}\right) = \frac{1}{a}\left(1 - \frac{b}{a}\right). \tag{2.4.8}$$

For $a = 20$, $b = 1$ we obtain:

$$\begin{aligned} \tfrac{1}{21} = (21)^{-1} = (20+1)^{-1} &= \tfrac{1}{20}(1 - \tfrac{1}{20}) \\ &= 0.05 \times 0.95 = 0.0475 \quad (0.04762). \end{aligned}$$

For $a = 20$, $b = -1$ we obtain:

$$\begin{aligned} \tfrac{1}{19} = (19)^{-1} = (20-1)^{-1} &= \tfrac{1}{20}[1 - (-1)\tfrac{1}{20}] \\ &= 0.05 \times 1.05 = 0.0525 \quad (0.05263). \end{aligned}$$

For $n = \frac{1}{2}$, (2.4.7) gives:

$$\sqrt{a+b} = \sqrt{a}\left(1 + \frac{1}{2}\frac{b}{a}\right). \tag{2.4.9}$$

For example, for $a = 16$, $b = 1$:

$$\begin{aligned} \sqrt{17} = \sqrt{16+1} = \sqrt{16}(1 + \tfrac{1}{2}\tfrac{1}{16}) &= 4(1 + \tfrac{1}{32}) \\ &= 4 \times 1.03 = 4.12 \quad (4.1231). \end{aligned}$$

Similarly, for $n = \frac{1}{3}$:

$$\sqrt[3]{a+b} = \sqrt[3]{a}\left(1 + \frac{1}{3}\frac{b}{a}\right), \tag{2.4.10}$$

so that:

$$\begin{aligned} \sqrt[3]{26} = \sqrt[3]{27-1} = \sqrt[3]{27}\left(1 + \frac{1}{3}\frac{-1}{27}\right) &= 3\left(1 - \frac{1}{81}\right) \\ &= 3 \times 0.988 = 2.964 \quad (2.9625). \end{aligned}$$

Evaluate the following numbers by the approximate formulas of this section. (The correct values are given in parentheses.)

2.4.1 $(201)^3 \simeq 8{,}120{,}000$ (8,120,601).

2.4.2 $\sqrt{98} \simeq 9.90$ (9.899).

2.4.3 $\sqrt[3]{1002} \simeq 10.0067$ (10.0067).

2.4.4 $\frac{1}{101} \simeq 0.0099 \quad (0.009901)$.

2.4.5 $\sqrt{0.99} \simeq 0.995 \quad (0.99499)$.

2.4.6 $\sqrt[3]{0.90} \simeq 0.967 \quad (0.96549)$.

2.4.7 $\sqrt[4]{1.10} \simeq 1.025 \quad (1.0242)$.

2.4.8 $\sqrt[6]{2.05} \simeq 1.12 \quad (1.1271)$.

2.4.9 $\frac{1}{\sqrt{104}} \simeq 0.098 \quad (0.09806)$.

2.4.10 $\sqrt[3]{26.7} \simeq 2.989 \quad (2.9888)$.

2.5 POLYNOMIAL EVALUATION

The cost of high-rise buildings consists of the sum of a variety of components, some of which are proportional to the number of floors n, others which grow as the square and as the cube of n, and others which are independent of n. Hence, indicating some constants with a, b, c, d, the cost C of a building may be expressed as:

$$C = an^3 + bn^2 + cn + d. \tag{2.5.1}$$

C is a *third degree polynomial* in n. As n varies, C changes and we would like to plot the cost C, as well as the cost per floor C/n, versus n. Let us measure C in thousands of dollars, assuming $a = 0.04$, $b = 2$, $c = 160$, $d = 2{,}000$:

$$C = 0.04n^3 + 2n^2 + 160n + 2{,}000. \tag{2.5.2}$$

The evaluation of polynomials is most efficiently performed by writing them in terms of "nested brackets" (see Section 2.2). In our example, the first two terms of (2.5.2) may be written, factoring out n^2, as:

$$0.04n^3 + 2n^2 = n^2(0.04n + 2),$$

the first three terms, by factoring out n, as:

$$n^2(0.04n + 2) + 160n = n[n(0.04n + 2) + 160],$$

and the entire polynomial as:

$$C = n[n(0.04n + 2) + 160] + 2{,}000. \tag{2.5.2a}$$

In this equivalent form C is rapidly evaluated by using the *constant multiplier* n as follows: (1) multiply 0.04 by n and add 2; (2) multiply this result by n and add 160; (3) multiply this result by n and add 2,000. The evaluation of (2.5.2a) entails 3 multiplications and 3 sums, while the evaluation of (2.5.2) would require 2 multiplications to obtain n^2 and n^3, 3 more to obtain the first 3 terms of (2.5.2) and 3 sums. *Synthetic substitution*, as the evaluation of (2.5.2a) is also called, thus saves 2 multiplications. Synthetic substitution is easily performed by the scheme of Tables 2.5.1 and 2.5.2, where $n = 10$ and $n = 20$, respectively.

Table 2.5.1

Multiplier	0.04	2+	160+	2,000+	
$n = 10$		$10 \times 0.04 = 0.4$	$10 \times 2.4 = 24$	$10 \times 184 = 1{,}840$	
	0.04	2.4	184	3,840	$= C$(thousands of \$)
					$\frac{C}{n} = 384$(thousands of \$)

Table 2.5.2

Multiplier	0.04	2+	160+	2,000+	
$n = 20$		$20 \times 0.04 = 0.8$	$20 \times 2.8 = 56$	$20 \times 216 = 4{,}320$	
	0.04	2.8	216	6,320	$= C$(thousands of \$)
					$\frac{C}{n} = 316$(thousands of \$)

Figure 2.1 gives C and C/n in terms of n and shows that the cost per floor is minimum at $n = 23$, where $C = 7{,}224$ and $C/n = 314$.

Try your hand at evaluating the following polynomials in x for the given values of x.

Note: If a power of x is lacking in the polynomial, set its coefficient equal to zero:

$$2x^3 + 3x + 2 = 2x^3 + 0x^2 + 3x + 2.$$

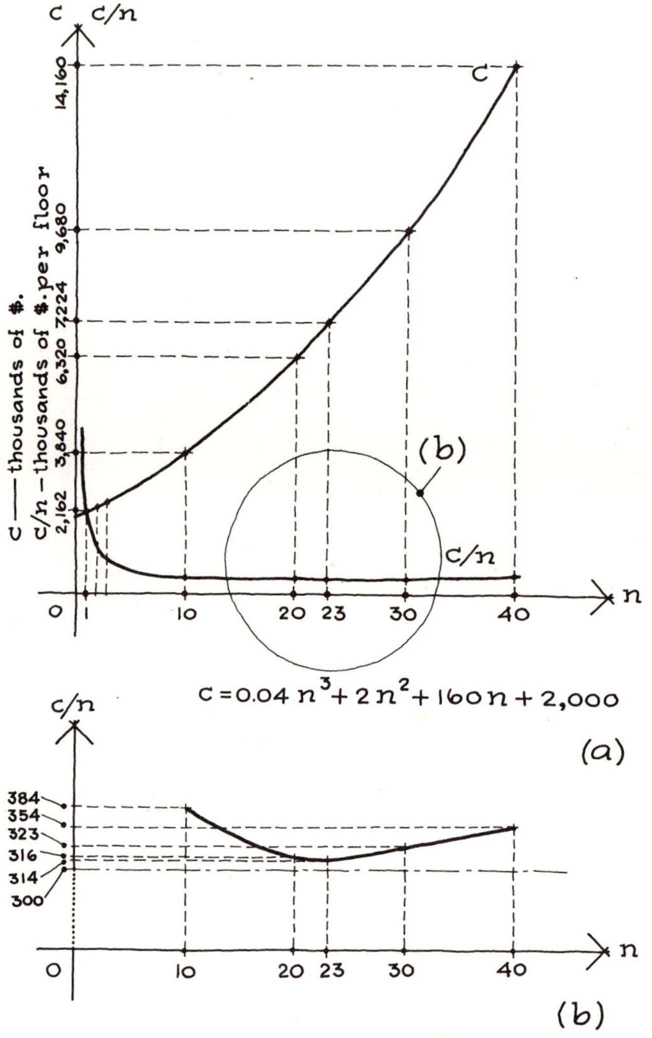

Figure 2.1

2.5.1 $P(x) = 2x^3 - x^2 + x + 1; \quad x = 2, -1.$
$P(2) = 15; \quad P(-1) = -3.$

2.5.2 $P(x) = x^3 - x + 2; \quad x = 2, -2.$
$P(2) = 8; \quad P(-2) = -4.$

2.5.3 $P(x) = 4x^4 - 3x^3 - 2x^2 + x - 1; \quad x = -1, +1.$
$P(-1) = 3; \quad P(+1) = -1.$

2.5.4 $P(x) = x^3 + 3x - 7;\quad x = -1, 1.$
$P(-1) = -11;\quad P(1) = -3.$

2.6 REAL ROOTS OF ALGEBRAIC EQUATIONS

Problems 2.5.1, 2.5.2 and 2.5.3 show that the corresponding polynomials $P(x)$ change sign (from plus to minus) as x varies, respectively, from $+2$ to -1, from $+2$ to -2, and from -1 to $+1$. The polynomial of problem 2.5.4, instead, stays negative for x varying from -1 to $+1$. If a polynomial changes sign for two values x_1, x_2 of x it *must* become zero for at least one value of x between x_1 and x_2. Such a value of x is called a *root* of the *algebraic equation* $P(x) = 0$. The roots of algebraic equations may be either real or complex numbers. We shall only be concerned here with real roots. Real roots of algebraic equations may be located by trial and error and will be shown in Section 5.5 to be essential in determining the solution of maximum and minimum problems.

For example, given the $P(x)$ of 2.5.1, we notice that $P(0) = +1$ and $P(-1) = -3$. Hence, a root of $P(x) = 0$ lies between $x = -1$ and $x = 0$. We try $x = -0.5$ by synthetic substitution:

Multiplier	2	−1	1	1	
$x = -0.5$		−1	1	−1	
	2	−2	2	0	$= P(-0.5)$

Since $P(-0.5) = 0$, $x = -0.5$ is an exact root of the *cubic equation*:

$$P(x) = 2x^3 - x^2 + x + 1 = 0.$$

In general, roots can only be evaluated approximately. For example, the polynomial of 2.5.3 is positive for $x = -1$ $[P(-1) = 3]$, negative for $x = 0$ $[P(0) = -1]$. We try $x = -0.5$ by synthetic substitution:

Multiplier	4	−3	−2	1	−1	
$x = -0.5$		−2	2.5	−0.25	−0.375	
	4	−5	0.5	0.75	−1.375	$= P(-0.5)$

We try $x = -0.75$:

Multiplier	4	-3	-2	1	-1	
$x = -0.75$		-3	4.5	-1.875	$+0.66$	
	4	-6	2.5	-0.875	-0.34	$= P(-0.75)$

We try $x = -0.8$:

Multiplier	4	-3	-2	1	-1	
$x = -0.8$		-3.2	4.96	-2.37	1.10	
	4	-6.2	2.96	-1.37	$+0.10$	$= P(-0.8)$

Since $P(-0.75) = -0.34 < 0$ and $P(-0.80) = 0.10 > 0$, the root x lies between 0.8 and 0.75. Moreover, since 0.10 is smaller than 0.34 the root x is probably nearer 0.80 than 0.75. [The method explained in Section 4.5 may be used to refine it to $x = -0.79$ with an error $P(-0.79) = 0.0033$.]

The roots x_1 and x_2 of a second degree or *quadratic equation:*

$$ax^2 + bx + c = 0 \tag{2.6.1}$$

can be obtained by formula:

$$x_{1,2} = \frac{1}{2a}(-b \pm \sqrt{b^2 - 4ac}). \tag{2.6.2}$$

There are also formulas for the solution of third- and fourth-degree equations, but equations of a degree higher than four *cannot* be solved by formula. They must be solved approximately.

As an example of problems leading to quadratic equations, consider the "break-even point" problem of determining the number of stories of a building for which monthly expenses and rents will balance. If the expenses E depend on some factors independent of the number of stories n and on others proportional to the square of n, and if the rents R are approximately proportional to n:

$$E = c + an^2; \qquad R = bn.$$

The break-even point is reached when $E = R$, that is, when:

$$c + an^2 = bn \qquad \therefore \quad an^2 - bn + c = 0.$$

For $a = 1$, $b = 25$, $c = 100$, we obtain:

$$n^2 - 25n + 100 = 0$$

and by (2.6.2):

$$\begin{aligned} n_{1,2} &= \frac{1}{2 \times 1}[-(-25) \pm \sqrt{(-25)^2 - 4 \times 1 \times 100}] \\ &= \tfrac{1}{2}(25 \pm \sqrt{625 - 400}) = \tfrac{1}{2}(25 \pm \sqrt{225}) \\ &= \tfrac{1}{2}(25 \pm 15); \qquad n_1 = 20,\ n_2 = 5. \end{aligned}$$

A building with either 20 or 5 stories will break even, but the higher building will be chosen if the land cost is an essential factor.

Apply the methods of this section to locate the real root of the following equations contained in the interval given in parentheses. Be satisfied with roots correct to two significant figures and check the results by substituting x in the equations.

2.6.1 $x^2 - 0.2x - 1.2 = 0$.
$(x_1 = 1.20,\ x_2 = -1.00)$

2.6.2 $x^3 - 1.4x^2 - 1.4x - 2.4 = 0$ (2, 3).

2.6.3 $x^3 - x^2 + 3.2x + 3.2 = 0$ $(-1, 0)$.

2.6.4 $x^4 - x - 1 = 0$ (1, 2).

2.6.5 $24x^3 - 2x + 3 = 0$ (find an inverval yourself).

2.6.6 $x^3 + x^2 - 1.44x + 1.44 = 0$ $(-1, -3)$.

2.7 SIMULTANEOUS LINEAR EQUATIONS

A new housing development may be built with 30 one-family houses and 10 two-family houses, or with 10 one-family houses and 20 two-family houses for the same cost of half a million dollars. What is the cost of a one-family house? A two-family house?

Calling x and y, respectively, the costs *in tens of thousands of dollars* of a one- and a two-family house, the first type of development will cost $30x + 10y$, the second $10x + 20y$. Since in both cases the cost is $\$500{,}000 = \$50 \times 10{,}000$:

$$\begin{aligned} &\text{I.} \quad & 30x + 10y &= 50 \\ &\text{II.} \quad & 10x + 20y &= 50. \end{aligned} \tag{2.7.1}$$

(2.7.1) is an example of a set of two *simultaneous linear equations.* Sets of simultaneous equations must be solved very often in architecture, since they govern all kinds of common problems, among them, most structural problems. The set of equations (2.7.1) is easily solved by elimination as follows. To eliminate y, divide equation I by 10 and equation II by 20:

$$\begin{aligned} &\text{I}'. \quad & 3x + y &= 5 \\ &\text{II}'. \quad & 0.5x + y &= 2.5. \end{aligned}$$

Now that y has the same coefficient in both equations, subtract II′ from I′:

$$2.5x + 0 = 2.5 \qquad \therefore \quad x = 1 \ (= \$10{,}000).$$

From I′, with $x = 1$:

$$y = 5 - 3x = 5 - 3 = 2 \ (= \$20{,}000).$$

The same process may be applied to a set of 3 or more equations. For example, given:

$$\begin{aligned} &\text{I.} \quad & 2x + 4y + 6z &= 18 \\ &\text{II.} \quad & 2x + 6y + 8z &= 24 \\ &\text{III.} \quad & 3x + 12y - 6z &= 3, \end{aligned}$$

divide I by 2, II by 2 and III by 3, to obtain equations with the same coefficient of x:

$$\begin{aligned} &\text{I}'. \quad & x + 2y + 3z &= 9, \\ &\text{II}'. \quad & x + 3y + 4z &= 12, \\ &\text{III}'. \quad & x + 4y - 2z &= 1. \end{aligned}$$

Subtract I′ from II′ and I′ from III′ to eliminate x:

$$\begin{aligned} &\text{I}''. \quad & 0 + y + z &= +3 \\ &\text{II}''. \quad & 0 + 2y - 5z &= -8. \end{aligned}$$

Divide II″ by 2:

$$\text{II}'''. \qquad y - 2.5z = -4,$$

and subtract II‴ from I″, obtaining:

$$0 + 3.5z = 7 \qquad \therefore \quad z = 2.$$

From I″:

$$y = 3 - z = 3 - 2 = 1.$$

From I′:

$$x = 9 - 2y - 3z = 9 - 2 \times 1 - 3 \times 2 = 1.$$

Systems of many equations must be solved by more systematic and efficient procedures, but elimination will do in elementary problems, provided *all* equations are checked at the end of the calculations. In our example, the original system I, II, III is satisfied, since:

$$\begin{aligned} 2 \times 1 + \ \ 4 \times 1 + 6 \times 2 &= 18, \\ 2 \times 1 + \ \ 6 \times 1 + 8 \times 2 &= 24, \\ 3 \times 1 + 12 \times 1 - 6 \times 2 &= \ \ 3. \end{aligned}$$

Electronic computers allow the solution of many hundreds and even thousands of equations and must be used in complex problems, which are not in the province of the architect.

Find the values of the unknowns x, y, z, etc., by elimination and *check your answers* by substituting the roots in the equations.

2.7.1 $2x + 3y = 17$
$x - 2y = 8$

2.7.2 $x + y + z = 6$
$x - y + z = 4$
$x - y - z = -2$

2.7.3 $1.1x + 3.5y = 11.33$
$2.2x + 4.6y = 17.38$

2.7.4 $0.5x + 1.0y + 1.5z = 7.0$
$1.0x - 0.5y + 2.0z = 6.0$
$2.0x - 1.0y + 4.0z = 12.0$

2.7.5 $0.05x + 1.12y = 3.72$
$1.79x + 0.07y = -1.10$

2.7.6
$$\begin{aligned} 1.2x + 1.3y + 2.5z + 2.7t &= 7.44 \\ -0.7x + 2.1y + 3.0z + 1.9t &= 5.63 \\ 1.6x + 0.4y + 27.0z - 1.8t &= 83.52 \\ 2.7x + 0.6y + 1.2z + 4.5t &= 7.35 \end{aligned}$$

The roots are: $x = 1.0$, $y = -2.2$, $z = 3.1$, $t = 0.5$. They are given so that you may check this more ambitious problem.

3

TRIGONOMETRY

3.1 INTRODUCTION

Trigonometry is a technique for dealing with triangles. Since most geometries can be "triangulated" and geometry is an essential component of architecture, you must be conversant with trigonometry. Mainly because it shortcuts long calculations, using a single standard table of numbers, trigonometry is a most efficient tool in quantitative geometrical work.

3.2 THE TRIGONOMETRIC FUNCTIONS

Draw a circle of radius R (Fig. 3.1) and a radial line OP making an angle α* with the horizontal OB. The angle α is measured moving against the hands of the clock, i.e., in the *counterclockwise direction*. From the

*α (alpha) is the Greek letter for a. The Greeks invented geometry and to this day all over the world angles are labeled by Greek letters, like α (alpha), β (beta), γ (gamma), θ (theta).

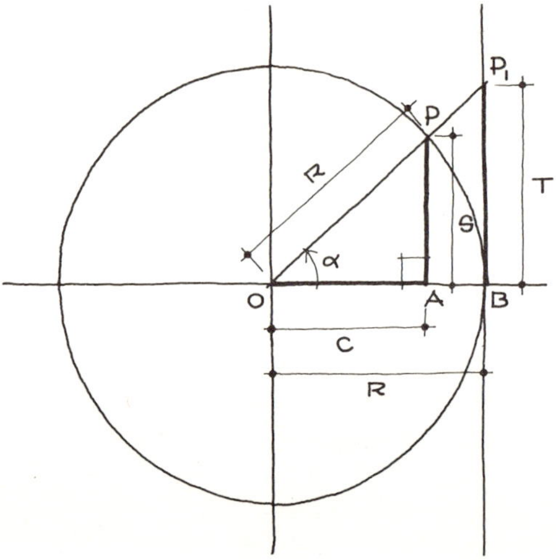

Figure 3.1

intersection P of this line with the circle draw the perpendicular PA to OB, thus obtaining the right triangle OPA.

Draw the vertical tangent to the circle at B and call P_1 the intersection of OP with this vertical line, thus obtaining another right triangle OP_1B.

The ratios of the lengths S, C and T of AP, OA and BP_1 to the length R of the radius occur so often in triangle calculations that they have been given names. Since the values of these ratios depend on the value of α, they are indicated by the shorthand symbols:

$$\frac{AP}{OP} = \sin\alpha, \quad \frac{OA}{OP} = \cos\alpha, \quad \frac{BP_1}{OP} = \frac{BP_1}{OB} = \tan\alpha,$$

or by:

$$\sin\alpha = \frac{S}{R}, \quad \cos\alpha = \frac{C}{R}, \quad \tan\alpha = \frac{T}{R}. \tag{3.2.1}$$

$\sin\alpha$ (read "sine alpha"), $\cos\alpha$ (read "cosine alpha") and $\tan\alpha$ (read "tangent alpha") are called the *trigonometric* or *circular functions.* Their values in terms of the angle α may be obtained directly from the drawing of Fig. 3.1, if R is chosen equal to 1 (say, 1 in.), since then $\sin\alpha = S/1 = S$, $\cos\alpha = C/1 = C$, $\tan\alpha = T/1 = T$. Thus, on the so-called *unit circle* the trigonometric functions are measured by the segments $AP = S$, $OA = C$, $BP_1 = T$.

The values of $\sin\alpha$, $\cos\alpha$, and $\tan\alpha$ for values of α between 0° and 45°* have been computed once and for all and appear in Table 2 of the Appendix. Strictly speaking, the column for $\tan\alpha$ is unnecessary, since the triangles OAP and OBP_1 are similar (they have parallel sides) and, as you can check on Fig. 3.1, $T/R = S/C$, so that one can also write the tangent of α as:

$$\tan\alpha = \frac{S}{C} = \frac{S/R}{C/R} = \frac{\sin\alpha}{\cos\alpha}. \tag{3.2.2}$$

But the ratio S/C occurs so often that it is convenient to have its values in a table.

The reciprocals of the sine, cosine, and tangent are often referred to by separate names and symbols, as follows:

$$\frac{1}{\sin\alpha} \equiv \operatorname{cosec}\alpha \quad \text{(read "cosecant alpha")},$$

$$\frac{1}{\cos\alpha} \equiv \sec\alpha \quad \text{(read "secant alpha")},$$

$$\frac{1}{\tan\alpha} \equiv \cot\alpha \quad \text{(read "cotangent alpha")}.$$

While $\cot\alpha$ is tabulated because it is of frequent use (as shown in Section 3.4), we will not use $\operatorname{cosec}\alpha$ or $\sec\alpha$ in this book.

3.3 RIGHT TRIANGLE GEOMETRY

A right triangle is completely defined by giving the lengths of two of its sides (Fig. 3.2), since by Pythagoras' theorem:

$$a^2 + b^2 = c^2$$

$$\therefore \quad c = \sqrt{a^2 + b^2},$$

$$a = \sqrt{c^2 - b^2},$$

$$b = \sqrt{c^2 - a^2}. \tag{3.3.1}$$

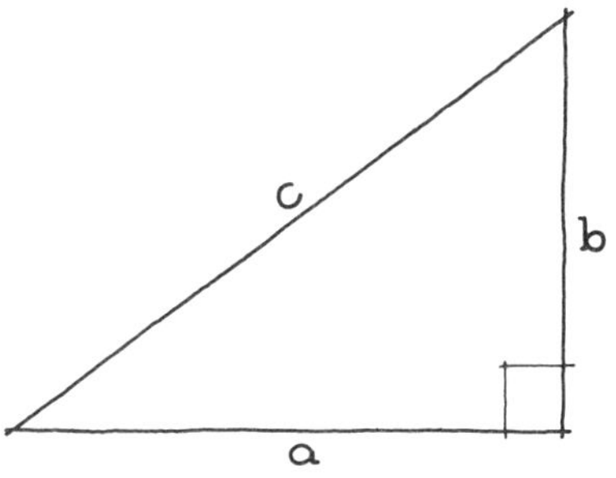

Figure 3.2

But a right triangle is also completely defined by giving the length of one of its sides and one of its angles (Fig.

*It will be seen in Section 3.5 that the values of the trigonometric functions for any value of α may be obtained from those for α varying from 0° to 45°.

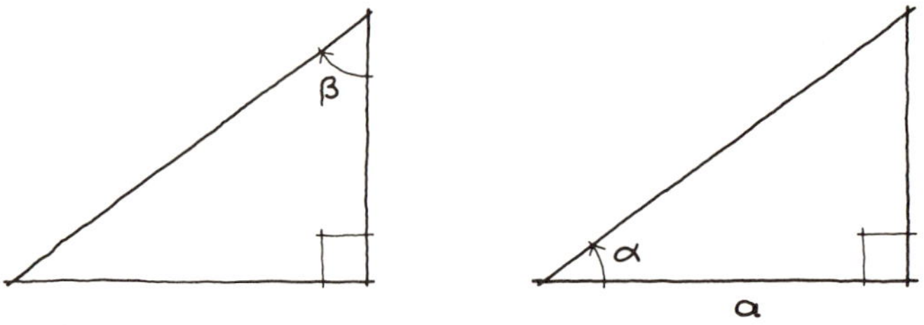

Figure 3.3

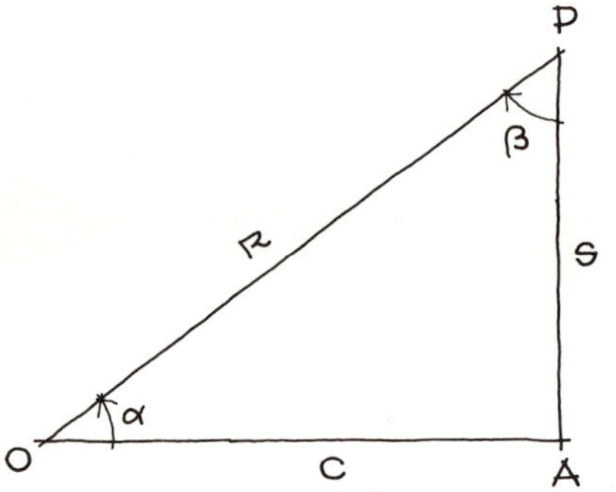

Figure 3.4

3.3). The other angle is known by the fact that the sum of the angles of a triangle always adds up to 180°, so that:

$$\alpha + \beta = 90^\circ$$
$$\therefore \quad \alpha = 90^\circ - \beta; \qquad (3.3.2)$$
$$\beta = 90^\circ - \alpha.$$

The other two sides are easily evaluated by a table of trigonometric functions. In fact, the first two equations (3.2.1) applied to the right triangle *OPA* of Fig. 3.1, which is reproduced in Fig. 3.4, give:

$$\sin\alpha = \frac{S}{R} \qquad \therefore \quad S = R\sin\alpha, \quad R = \frac{S}{\sin\alpha}, \qquad (3.3.3)$$

$$\cos\alpha = \frac{C}{R} \qquad \therefore \quad C = R\cos\alpha, \quad R = \frac{C}{\cos\alpha}, \qquad (3.3.4)$$

and (3.2.2) gives:

$$\tan\alpha = \frac{S}{C} \qquad \therefore \quad S = C\tan\alpha, \quad C = \frac{S}{\tan\alpha}. \qquad (3.3.5)$$

It is now clear that if one side and one angle of a right triangle are known, the most efficient way of obtaining the other sides is by trigonometry. The following examples will show you the practical use of the equations of this section.

(a) A ladder 20 ft long must not be inclined by less than 60° to the horizontal lest it should slide on the floor (Fig. 3.5). How many feet away from the wall can you set the foot of the ladder?

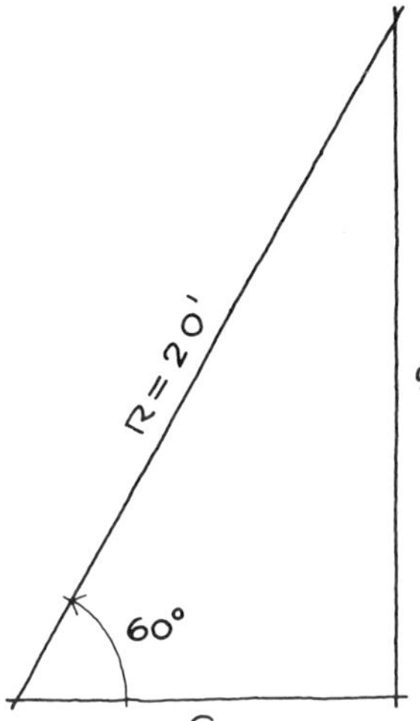

Figure 3.5

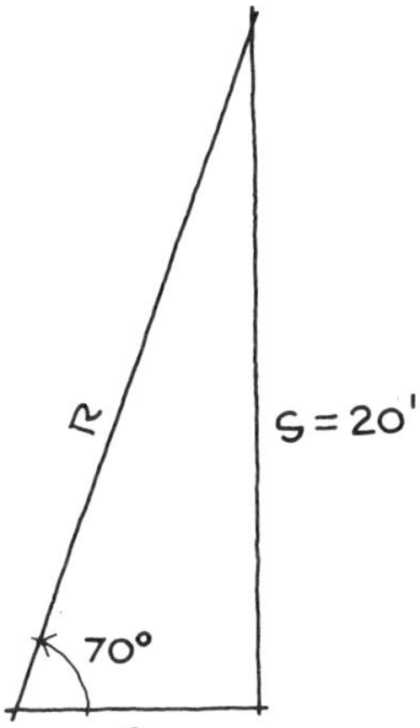

Figure 3.6

By (3.3.4) and Table 2:

$$C = R \cos \alpha = R \cos 60^\circ = 20 \times 0.5 = 10 \text{ ft.}$$

(b) How high does the ladder reach?
By (3.3.3) and Table 2:

$$S = R \sin \alpha = 20 \sin 60^\circ = 20 \times 0.866 = 17.32 \text{ ft} = 17 \text{ ft-}3.8 \text{ in.}$$

(c) In order not to tip backwards when loaded, a ladder should not be inclined more than 70° to the horizontal (Fig. 3.6). What is the minimum length of a ladder reaching 20 ft?
By (3.3.3):

$$R = \frac{S}{\sin \alpha} = \frac{20}{\sin 70^\circ} = \frac{20}{0.94} = 21.3 \text{ ft} = 21 \text{ ft-}3.6 \text{ in.}$$

(d) How far is the foot of the ladder from the wall?
By (3.3.5):

$$C = \frac{S}{\tan \alpha} = \frac{20}{\tan 70^\circ} = \frac{20}{2.75} = 7.28 \text{ ft} = 7 \text{ ft-}3.4 \text{ in.}$$

(e) A garage ramp is 50 ft long in plan and slopes by 10° (Fig. 3.7). How high does it rise?
By (3.3.5):

$$S = C \tan \alpha = 50 \tan 10^\circ = 50 \times 0.176 = 8.8 \text{ ft} = 8 \text{ ft-}9.6 \text{ in.}$$

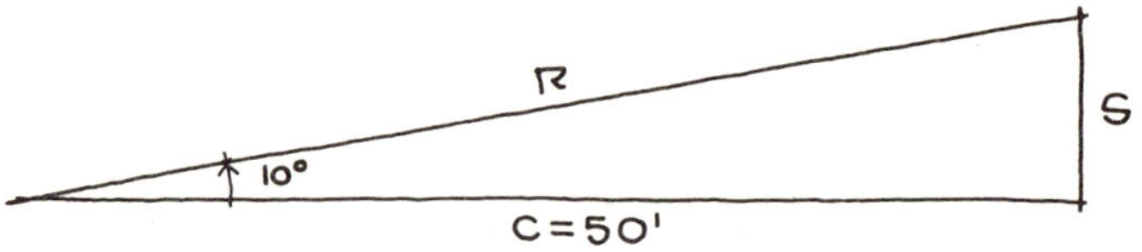

Figure 3.7

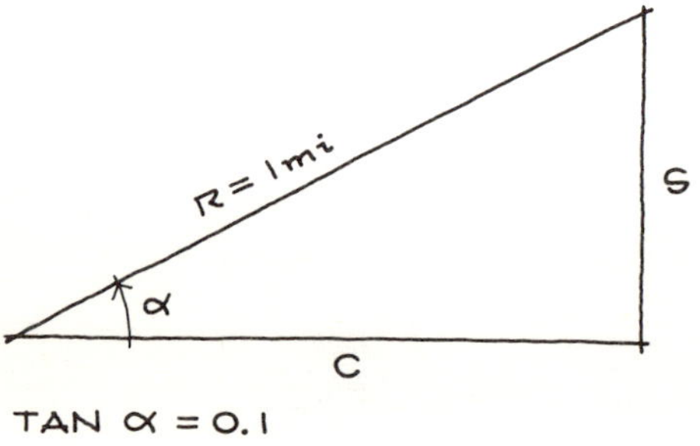

Figure 3.8

(f) A road has a constant slope of 1 in 10. How much does it rise in 1 mile? The slope is the tangent of the angle α (Fig. 3.8):

$$\tan \alpha = 0.1.$$

From Table 2, if $\tan \alpha = 0.1$, the angle α is 5°43′. The rise S is given by (3.3.3), with $\sin \alpha = \sin 5°43' = 0.0996$:

$$S = R \sin \alpha = 5{,}280 \times 0.0996 = 52.56 \text{ ft} = 52 \text{ ft-}6.7 \text{ in.}$$

3.4 TRIGONOMETRIC IDENTITIES

Trigonometric identities are relationships between the trigonometric functions of the same angle α, or of two angles α and β, which are true *whatever* the values of α and β. They allow the evaluation of the trigonometric functions *for any value of the angle* α *by means of the values of the functions for* α *between* 0° *and* 45° *given in Table* 2.
For example, since:

$$\sin \alpha = \frac{S}{R}, \qquad \cos \alpha = \frac{C}{R},$$

by Pythagoras' theorem:

$$\sin^2 \alpha + \cos^2 \alpha = \left(\frac{S}{R}\right)^2 + \left(\frac{C}{R}\right)^2 = \frac{S^2 + C^2}{R^2} = \frac{R^2}{R^2} = 1.^*$$

By this result, any one trigonometric function may be computed in terms

*$\cos^2 \alpha$, $\sin^2 \alpha$ and $\tan^2 \alpha$ are shorthand symbols for $(\cos \alpha)^2$, $(\sin \alpha)^2$, $(\tan \alpha)^2$.

of any other, as follows:
From:

$$\sin^2\alpha + \cos^2\alpha = 1, \tag{3.4.1}$$

we obtain:

$$\sin\alpha = \sqrt{1-\cos^2\alpha}, \qquad \tan\alpha = \frac{\sin\alpha}{\cos\alpha} = \frac{\sqrt{1-\cos^2\alpha}}{\cos\alpha}, \tag{3.4.2}$$

$$\cos\alpha = \sqrt{1-\sin^2\alpha}, \qquad \tan\alpha = \frac{\sin\alpha}{\sqrt{1-\sin^2\alpha}}. \tag{3.4.3}$$

From (3.4.2):

$$\tan^2\alpha = \frac{1-\cos^2\alpha}{\cos^2\alpha} \qquad \therefore \quad \cos^2\alpha\tan^2\alpha = 1-\cos^2\alpha,$$

$$\cos^2\alpha(1+\tan^2\alpha) = 1, \qquad \cos^2\alpha = \frac{1}{1+\tan^2\alpha},$$

$$\cos\alpha = \frac{1}{\sqrt{1+\tan^2\alpha}}. \tag{3.4.4}$$

From (3.4.3):

$$\tan^2\alpha = \frac{\sin^2\alpha}{1-\sin^2\alpha} \qquad \therefore \quad (1-\sin^2\alpha)\tan^2\alpha = \sin^2\alpha,$$

$$\sin^2\alpha(1+\tan^2\alpha) = \tan^2\alpha, \qquad \sin^2\alpha = \frac{\tan^2\alpha}{1+\tan^2\alpha},$$

$$\sin\alpha = \frac{\tan\alpha}{\sqrt{1+\tan^2\alpha}}. \tag{3.4.5}$$

Equations (3.4.1) to (3.4.5) are *identities*, i.e., are valid for any value of α. For example, given $\alpha = 30°$, $\sin\alpha = 0.5$, you may check by (3.4.3) that:

$$\cos\alpha = \sqrt{1-(0.5)^2} = \sqrt{1-0.25} = \sqrt{0.75} = 0.866,$$

$$\tan\alpha = \frac{0.5}{0.866} = 0.577,$$

as shown in Table 2 for $\alpha = 30°$.

To extend Table 2 to angles between 45° and 90°, consider the triangles of Figs. 3.9a and 3.9b, which are identical but rotated with respect to each other. From the first:

$$\sin\alpha = \frac{a}{c}, \quad \cos\alpha = \frac{b}{c}, \quad \tan\alpha = \frac{a}{b};$$

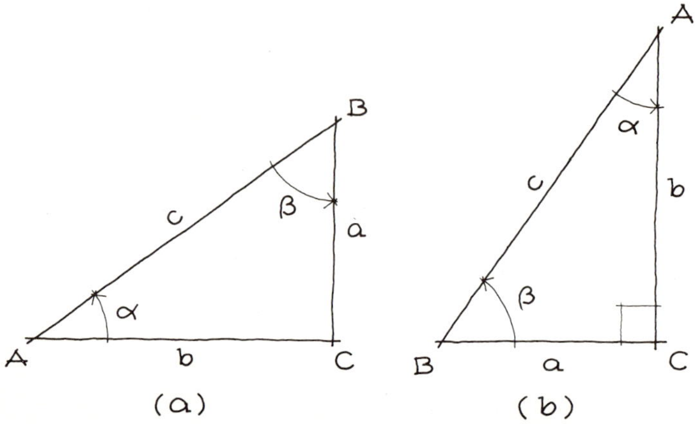

Figure 3.9

from the second:

$$\sin \beta = \frac{b}{c}, \quad \cos \beta = \frac{a}{c}, \quad \tan \beta = \frac{b}{a}.$$

Since $\beta = 90° - \alpha$:

$$\sin (90° - \alpha) = \cos \alpha, \qquad \cos (90° - \alpha) = \sin \alpha,$$
$$\tan (90° - \alpha) = \frac{1}{\tan \alpha}. \tag{3.4.6}$$

Check, for example, in Table 2 that $\sin 30° = 0.5 = \cos 60°$.

When two angles α, β are such that $\alpha + \beta = 90°$, they are called *complementary*. The *co*sine is the sine of the *co*mplementary angle. (3.4.6) shows why Table 2 can be limited to angles between 0° and 45° and may be read upwards for angles between 45° and 90°, as far as the sine and the cosine are concerned.

The function:

$$\cot \alpha \equiv \frac{1}{\tan \alpha} \text{ (read "cotangent } \alpha\text{")} \tag{3.4.7}$$

is listed to obtain $\tan (90° - \alpha)$ directly rather than by evaluating $1/\tan \alpha$ (*co*tangent is the tangent of the *co*mplementary angle). You may check in Table 2 that:

$$\tan 60° = \tan (90° - 30°) = \cot 30° = \frac{1}{\tan 30°} = \frac{1}{0.577} = 1.73.$$

Other identities, of greatest practical value, relate the trigonometric functions of the *sum* of two angles, $\alpha + \beta$, to the trigonometric functions of α and of β. They can be obtained by means of the unit circle of Fig. 3.10, in which the angle $\alpha = P_1OB$ is repeated following the angle $\beta = P_2OB$, so that:

$$P_3OP_2 = \alpha; \qquad P_3OB = \alpha + \beta.$$

Drawing from P_3 the perpendicular P_3A_1' to OP_2:

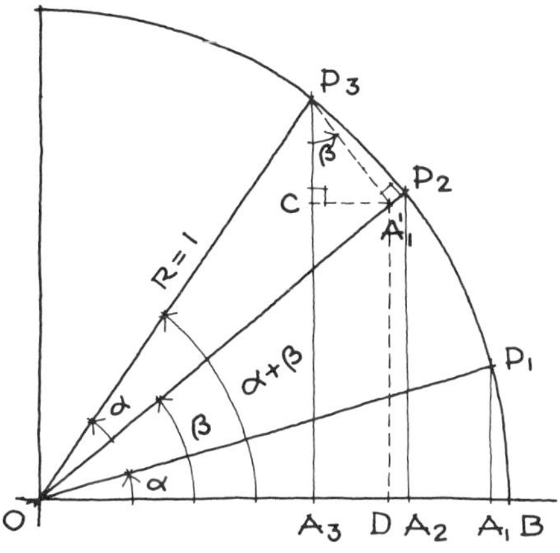

Figure 3.10

$$\sin\alpha = A_1P_1 = A_1'P_3, \qquad \cos\alpha = OA_1 = OA_1', \tag{a}$$

while by the definition of sine and cosine:

$$\sin\beta = A_2P_2, \qquad \cos\beta = OA_2,$$
$$\sin(\alpha+\beta) = A_3P_3, \qquad \cos(\alpha+\beta) = OA_3.$$

Draw the perpendiculars $A_1'C$ to P_3A_3 and $A_1'D$ to OB. Then:

$$\sin(\alpha+\beta) = A_3P_3 = A_3C + CP_3 = DA_1' + CP_3. \tag{b}$$

In the right triangle P_3CA_1' the angle at P_3 equals β; hence:

$$CP_3 = A_1'P_3\cos\beta = \sin\alpha\cos\beta. \tag{c}$$

In the right triangle $A_1'OD$:

$$DA_1' = OA_1'\sin\beta = \cos\alpha\sin\beta. \tag{d}$$

Hence, substituting (c) and (d) in equation (b):

$$\sin(\alpha+\beta) = \sin\alpha\cos\beta + \cos\alpha\sin\beta. \tag{3.4.8}$$

Similarly:

$$\cos(\alpha+\beta) = OA_3 = OD - DA_3 = OD - A_1'C$$
$$= OA_1'\cos\beta - A_1'P_3\sin\beta$$

and by equation (a):

$$\cos(\alpha + \beta) = \cos\alpha\cos\beta - \sin\alpha\sin\beta. \tag{3.4.9}$$

By means of (3.4.8), (3.4.9) we finally obtain:

$$\tan(\alpha + \beta) = \frac{\sin(\alpha + \beta)}{\cos(\alpha + \beta)} = \frac{\sin\alpha\cos\beta + \cos\alpha\sin\beta}{\cos\alpha\cos\beta - \sin\alpha\sin\beta},$$

or, dividing numerator and denominator by $\cos\alpha\cos\beta$:

$$\tan(\alpha + \beta) = \frac{\sin\alpha/\cos\alpha \times 1 + 1 \times \sin\beta/\cos\beta}{1 - (\sin\alpha/\cos\alpha)(\sin\beta/\cos\beta)};$$

that is:

$$\tan(\alpha + \beta) = \frac{\tan\alpha + \tan\beta}{1 - \tan\alpha\tan\beta}. \tag{3.4.10}$$

For the particular case $\beta = \alpha$, (3.4.8), (3.4.9), and (3.4.10) give:

$$\begin{aligned} \sin 2\alpha &= 2\sin\alpha\cos\alpha, \\ \cos 2\alpha &= \cos^2\alpha - \sin^2\alpha \\ \tan 2\alpha &= \frac{2\tan\alpha}{1 - \tan^2\alpha}. \end{aligned} \tag{3.4.11}$$

For example, you may check that:

$$\begin{aligned} \sin 90° &= \sin(30° + 60°) = \sin 30°\cos 60° + \cos 30°\sin 60° \\ &= (0.5 \times 0.5) + (0.866 \times 0.866) = 1. \end{aligned}$$

Obtain the values of the following trigonometric functions by the identities of this section and check approximately on a unit circle that they are correct. Check also by means of Table 2.

3.4.1 $\tan\alpha = 0.325$; $\cos\alpha =$, $\sin\alpha =$.

3.4.2 $\sin\alpha = 0.707$; $\cos\alpha =$, $\tan\alpha =$.

3.4.3 $\tan\alpha = 0.325$; $\tan(90° - \alpha) =$.

3.4.4 $\alpha = 10°$, $\beta = 20°$; $\cos(\alpha + \beta) =$.

3.4.5 $\alpha = 15°$, $\beta = 30°$; $\sin(\alpha + \beta) =$, $\cos(\alpha + \beta) =$, $\tan(\alpha + \beta) =$

3.4.6 $\alpha = 45°$, $\beta = 45°$; $\cos(\alpha + \beta) =$, $\sin(\alpha + \beta) =$
$\tan(\alpha + \beta) =$

3.5 THE FOUR QUADRANTS

In order to evaluate by means of Table 2 the trigonometric functions for angles in the 3 quadrants II, III, and IV of Fig. 3.11 in terms of those in the first quadrant I, one may use the identities of Section 3.4. But it is actually easier to visualize such relationships on a unit circle.

To this purpose let us measure the sine segment S positively above the horizontal OB and negatively below, and the cosine segment C positively to the right of O and negatively to the left of O (Fig. 3.12).

Figure 3.12 shows that in:

I. S is positive, C is positive, $T = S/C$ is positive.
II. S is positive, C is negative, $T = S/C$ is negative.
III. S is negative, C is negative, $T = S/C$ is positive.
IV. S is negative, C is positive, $T = S/C$ is negative.

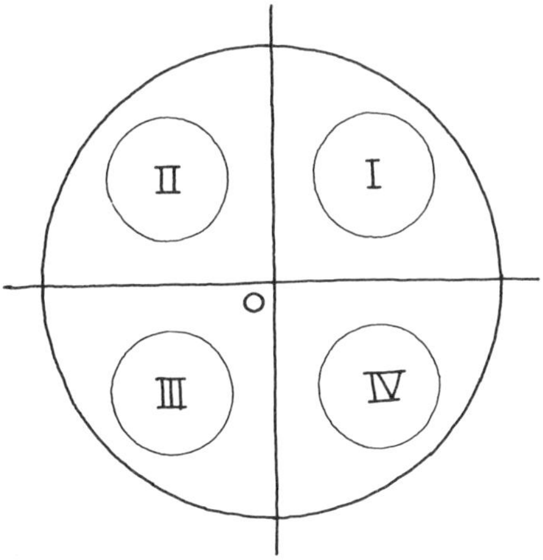

Figure 3.11

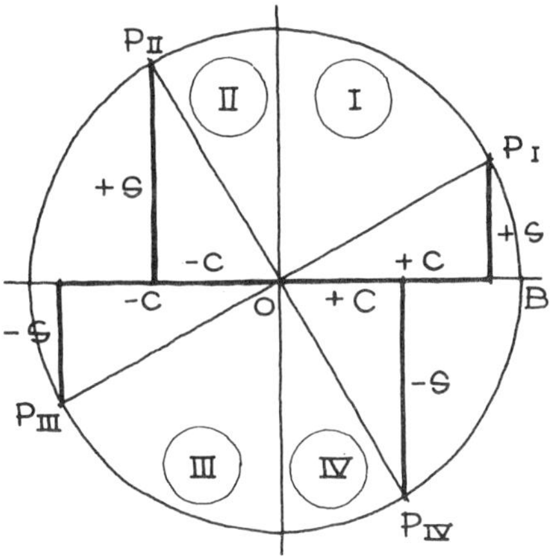

Figure 3.12

To obtain the trigonometric functions of an angle $\beta = \alpha + 90°$ in quadrant II, where α is in quadrant I, we inspect Fig. 3.13 and notice that $S_\beta = C_\alpha$, while $C_\beta = -S_\alpha$ since C_β is to the left of O (negative) and S_α is

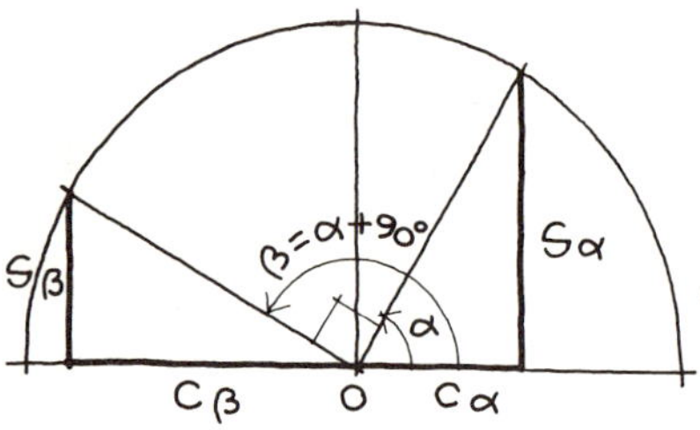

Figure 3.13

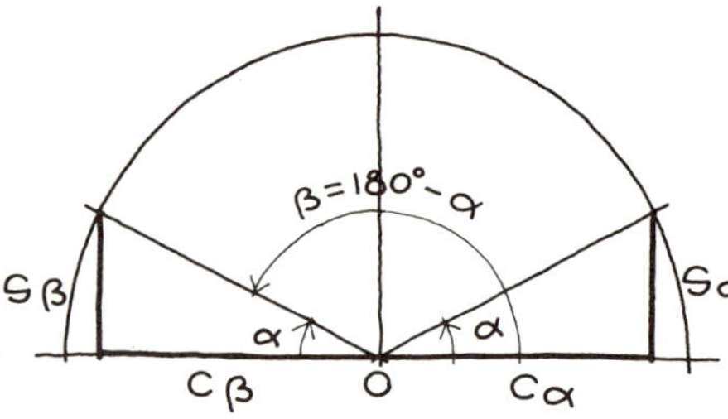

Figure 3.14

above OB (positive). Hence:

$$\sin(\alpha + 90^\circ) = \cos\alpha, \qquad \cos(\alpha + 90^\circ) = -\sin\alpha,$$

$$\tan(\alpha + 90^\circ) = \frac{\cos\alpha}{-\sin\alpha} = -\frac{1}{\sin\alpha/\cos\alpha} = -\frac{1}{\tan\alpha}. \tag{3.5.1}$$

Similarly, from Fig. 3.14, $S_\beta = S_\alpha$, $C_\beta = -C_\alpha$, or:

$$\sin(180^\circ - \alpha) = \sin\alpha, \qquad \cos(180^\circ - \alpha) = -\cos\alpha,$$

$$\tan(180^\circ - \alpha) = \frac{\sin\alpha}{-\cos\alpha} = -\tan\alpha. \tag{3.5.2}$$

From Fig. 3.15, $S_\beta = -S_\alpha$, $C_\beta = -C_\alpha$, or:

$$\sin(180^\circ + \alpha) = -\sin\alpha, \qquad \cos(180^\circ + \alpha) = -\cos\alpha,$$

$$\tan(180^\circ + \alpha) = \frac{-\sin\alpha}{-\cos\alpha} = \tan\alpha. \tag{3.5.3}$$

And from Fig. 3.16, $S_\beta = -S_\alpha$, $C_\beta = C_\alpha$, or:

$$\sin(360^\circ - \alpha) = \sin(-\alpha) = -\sin\alpha,$$

$$\cos(360^\circ - \alpha) = \cos(-\alpha) = \cos\alpha, \tag{3.5.4}$$

$$\tan(360^\circ - \alpha) = \tan(-\alpha) = \frac{-\sin\alpha}{\cos\alpha} = -\tan\alpha.$$

Since $360°—\alpha$ is equivalent to an angle equal to α in the *clockwise* direction, i.e., to an angle $-\alpha$, the identities (3.5.4) prove that if α is changed into $-\alpha$, the sine and the tangent change sign but not value, and the cosine remains unchanged:

$$\sin(-\alpha) = -\sin\alpha, \quad \cos(-\alpha) = \cos\alpha, \quad \tan(-\alpha) = -\tan\alpha. \tag{3.5.5}$$

Hence, changing the sign of β in (3.4.8), (3.4.9), (3.4.10), we obtain

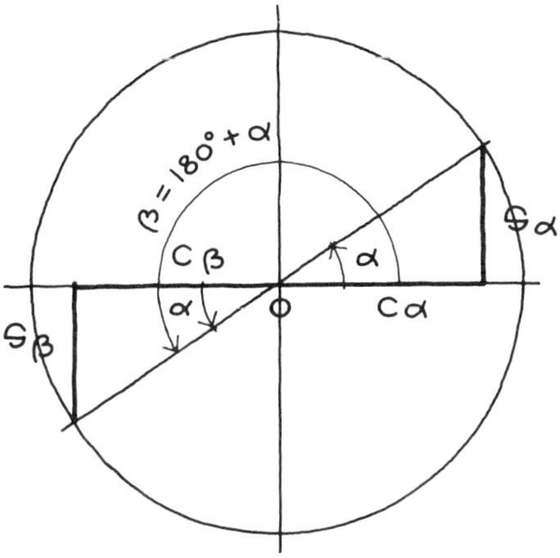

Figure 3.15

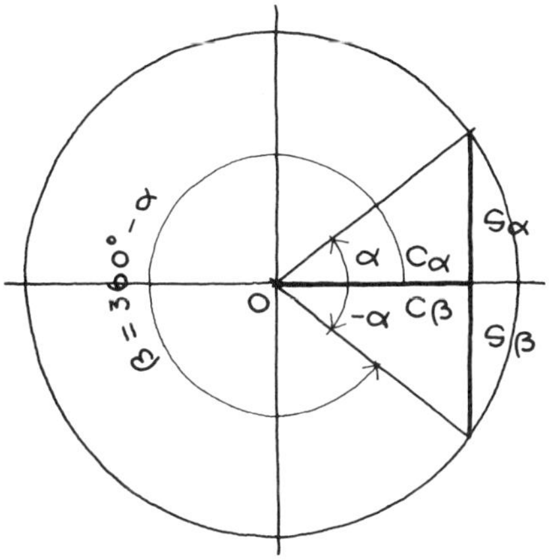

Figure 3.16

with $\sin(-\beta) = -\sin\beta$, $\cos(-\beta) = \cos\beta$, $\tan(-\beta) = -\tan\beta$:

$$\sin(\alpha - \beta) = \sin\alpha\cos\beta - \cos\alpha\sin\beta,$$
$$\cos(\alpha - \beta) = \cos\alpha\cos\beta + \sin\alpha\sin\beta, \qquad (3.5.6)$$
$$\tan(\alpha - \beta) = \frac{\tan\alpha - \tan\beta}{1 + \tan\alpha\tan\beta}.$$

You may now evaluate by Table 2 the following trigonometric functions.

3.5.1 $\cos 120° = -0.5$; $\sin 120° = 0.866$; $\tan 120° = -1.73$.

3.5.2 $\sin(-45°) = -0.707$; $\cos(-45°) = 0.707$; $\tan(-45°) = -1$.

3.5.3 $\sin 216° = -0.588$; $\cos 216° = -0.809$; $\tan 216° = 0.727$.

3.5.4 $\sin 320° = -0.643$; $\cos 320° = 0.766$; $\tan 320° = -0.839$.

3.5.5 $\sin(-18°) = -0.309$; $\cos(-18°) = 0.951$; $\tan(-18°) = -0.325$.

3.6 RADIAN MEASURE OF ANGLES

Given an angle α in degrees, you may draw by means of a protractor the line OP defining the angle α on the unit circle and measure on the unit circle the length x of the arc BP (Fig. 3.17). On the other hand, you may obtain the angle α in degrees as soon as the length x of the arc BP (from B

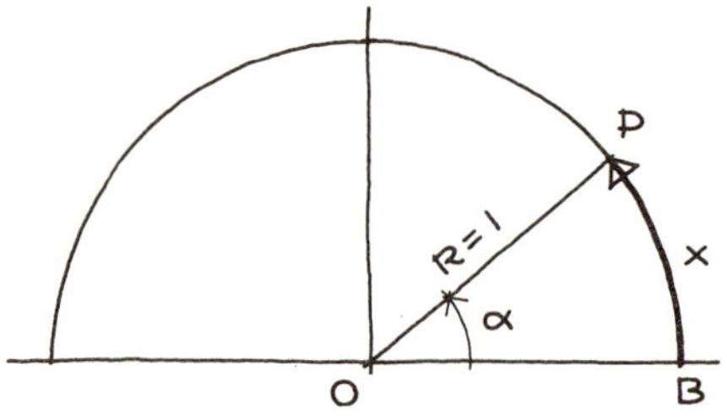

Figure 3.17

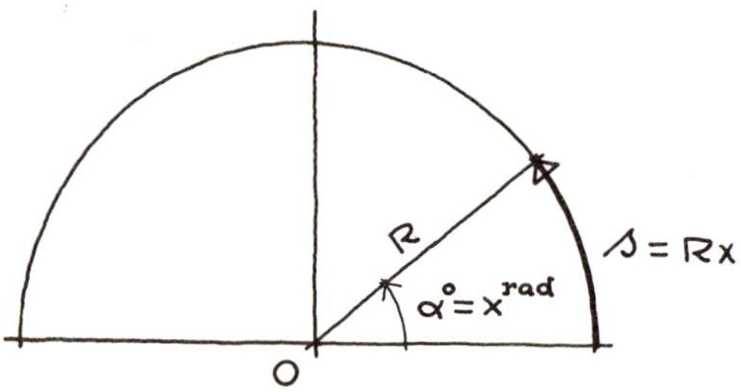

Figure 3.18

towards P) on the unit circle is given, since knowing x you may plot P, draw OP and measure α by means of a protractor. Hence either α or x can be used to measure an angle. The length x on the unit circle is called the *measure of the angle* α *in radians* (abbreviated *rad*) and is used very often in trigonometry, particularly when trigonometry is introduced in problems involving the calculus.

The radian measure of an angle, just as its degree measure, has no units, since it is the ratio of the length x of the arc to the length of the radius, which in this case is 1. For example, since the circumference of the circle equals $2\pi R$, the length x on the unit circle ($R = 1$) for an angle $\alpha = 360°$ is $2\pi \times 1 = 2\pi$ and:

$$360° = 2\pi \text{ rad} = 2 \times 3.14 \text{ rad} = 6.28 \text{ rad}. \qquad (3.6.1)$$

Hence:

$$270° = \tfrac{3}{4}\, 360° = \tfrac{3}{4}\, 2\pi \text{ rad} = \tfrac{3}{2}\, \pi \text{ rad} = 4.71 \text{ rad}.$$

$$180° = \frac{360°}{2} = \frac{2\pi}{2} \text{ rad} = \pi \text{ rad} = 3.14 \text{ rad}.$$

$$90° = \frac{360°}{4} = \frac{2\pi}{4} \text{ rad} = \frac{\pi}{2} \text{ rad} = 1.57 \text{ rad}.$$

Since the length s of a circular arc is the product of the radius R and the angle it subtends (Fig. 3.18) and since x is the measure of the angle for $R = 1$, the length s of a circular arc of radius R and angle x rad is given by:

$$\frac{s}{R} = \frac{x}{1} \qquad \therefore \quad s = Rx. \qquad (3.6.2)$$

From (3.6.1), dividing both sides by $2\pi = 6.28$:

$$\frac{360°}{6.28} = 57.3° = 57°17' = \frac{2\pi}{2\pi} = 1 \text{ rad}. \qquad (3.6.3)$$

Hence, to change an angle measure from degrees to radians you must divide $\alpha°$ by 57.3, and to change an angle measure from radians to degrees you must multiply x rad by 57.3.

3.6.1 A bicycle wheel has a radius of 1 ft and spokes spaced by 1.5 in. on its rim. What is the angle between the spokes?

With $s = 1.5$ in. $= 0.125$ ft and $R = 1$ ft, (3.6.2) gives:

$$x = \frac{0.125}{1} = 0.125 \text{ rad}; \ \alpha = 0.125 \times 57.3° = 7.16° = 7°9.6'.*$$

3.6.2 A clock has a minute hand 2 ft long. How long is the arc described by the hand tip in 10 minutes?

Ten minutes is $\frac{1}{6}$ of an hour. The minute hand covers in 10 minutes an angle $\alpha = (\frac{1}{6})\, 360°$ or $x = (\frac{1}{6})\, 2\pi = (\pi/3) = (3.14/3) = 1.05$ rad. By (3.6.2) the path described by its tip has a length:

$$s = Rx = 2 \times 1.05 = 2.10 \text{ ft.}$$

Check by means of Table 2 the values of the given trigonometric functions for the values of angles given in radians.

3.6.3 $x = 0.314$ rad, $\sin x = 0.309$.

3.6.4 $x = 1.239$ rad, $\cos x = 0.326$.

3.6.5 $x = 2.79$ rad, $\cos x = -0.940$.

3.6.6 $x = 3.30$ rad, $\sin x = -0.156$.

3.6.7 $x = 4.38$ rad, $\cos x = -0.326$.

3.6.8 $x = 5.20$ rad, $\sin x = -0.883$.

3.6.9 $x = 7.00$ rad, $\tan x = 0.869$.
Hint: Subtract $360° = 2\pi$ rad.

3.6.10 $x = 21.10$ rad, $\sin x = 0.777$, $\cos x = -0.629$, $\tan x = -0.810$.
Hint: Subtract an integer number of times $360° = 2\pi$ until you obtain an angle less than 360°.

*To transform 0.16° into minutes of a degree, remember that $0.16 = 16/100$ and that 1 min $= 1°/60$:

$$\frac{x'}{60} = \frac{16}{100} \quad \therefore \quad x' = \frac{16 \times 60}{100} = 9.6'.$$

3.7 TRIGONOMETRIC FUNCTIONS OF SMALL ANGLES

The values of sin x, cos x and tan x may be evaluated approximately without using trigonometric tables, if x is much smaller than 1 ($x \ll 1$ rad). These approximate values of the trigonometric functions are essential in the applications of trigonometry to the calculus. It is seen from the unit circle of Fig. 3.19 that as x becomes very small, say $x = x'$, the segments $A_1'P' = \sin x'$ and $BP_1' = \tan x'$ become practically equal to the arc $BP' = x'$, and $OA_1' = \cos x'$ becomes practically equal to $OB = 1$. Hence:

$$\sin x \simeq x, \quad \tan x \simeq x, \quad \cos x \simeq 1 \qquad \text{for } x \ll 1, \tag{3.7.1}$$

where x is measured in radians.

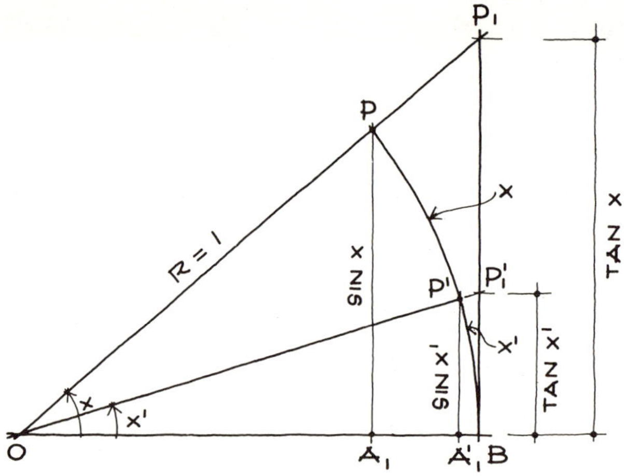

Figure 3.19

3.7.1 Check on Table 2 that the approximation (3.7.1) for the sine holds up to $\alpha = 7°$ to 3 decimal places and up to $\alpha = 19°$ to 2 decimal places.

3.7.2 Check the limits of validity of the approximations (3.7.1) for the cosine and the tangent to 2 decimal places. (5°, 17°.)

A better approximation of cos x may be obtained from Fig. 3.20. When x is much less than 1 rad the length of the chord BP may be approximated by the length x of the arc. The chord BP makes an angle $x/2$ with PA. Hence, approximating $PC = CB = \sin x/2$ by $x/2$ according to (3.7.1):

$$AB = PB \sin \frac{x}{2} \simeq x \times \frac{x}{2} = \frac{x^2}{2} = OB - OA = 1 - \cos x,$$

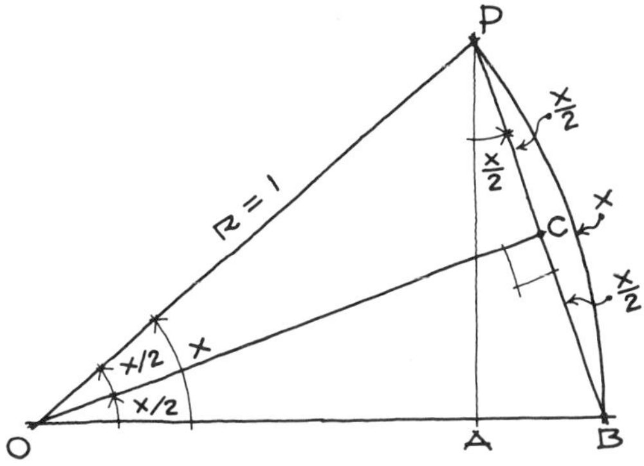

Figure 3.20

from which:

$$\cos x = 1 - \frac{x^2}{2}. \tag{3.7.2}$$

The corresponding approximations for sin x and tan x may be shown to be:

$$\sin x = x - \frac{x^3}{6}, \tag{3.7.3}$$

$$\tan x = x + \frac{x^3}{3}. \tag{3.7.4}$$

3.7.3 Check that (3.7.3) holds up to $\alpha = 34°$ to 3 decimal places.

3.7.4 Check the limits of validity to 2 decimal places of (3.7.2), (3.7.3) and (3.7.4). (29°, 44°, 32°.)

4

ANALYTIC GEOMETRY

4.1 INTRODUCTION

You have seen in Chapters 2 and 3 the roles algebra and geometry can play in our work. Geometry allows the visualization of problems; algebra simplifies the quantitative evaluation of their answers.

By "marrying geometry to algebra" Descartes, the inventor of analytic geometry, has shown that geometry and algebra are two ways of looking at the same problem. Analytic geometry allows us to visualize algebra and to "algebrize" geometry. Those of you who are "geometrically minded" will find analytic geometry most helpful in solving algebraic problems visually. Those of you who are "numerically minded" will find it ideal to obtain quick and accurate answers to geometrical problems.

4.2 POINTS

A point P can be located in a plane by giving the distances of the point from 2 intersecting lines. To this purpose, we choose two basic lines, respectively horizontal and vertical, called the x-axis and the y-axis or, also, the *axis*

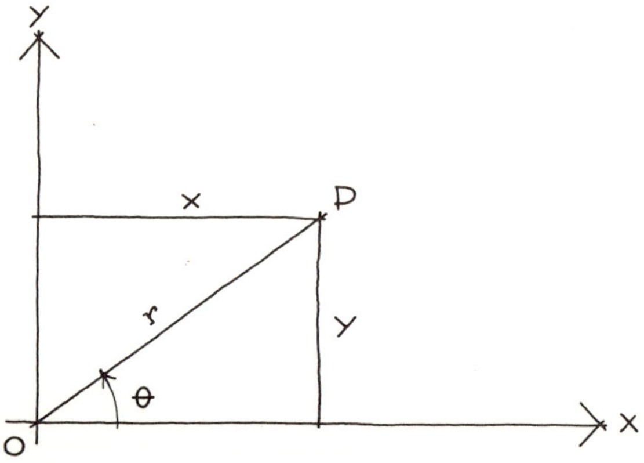

Figure 4.1

of the abscissas and the *axis of the ordinates*, which meet at a point O, called the *origin of the coordinate axes* (Fig. 4.1). The abscissas x are measured positively to the right of O, negatively to the left of O. The ordinates y are measured positively above and negatively below O.

The distance r of P from O, by Pythagoras' theorem, equals:

$$OP = r = \sqrt{x^2 + y^2}. \tag{4.2.1}$$

By trigonometry, the angle θ (read "theta") between the positive x-axis and the line OP is such that:

$$\tan \theta = \frac{y}{x}. \tag{4.2.2}$$

Thus, y/x is the slope of the line OP. Figure 4.2 shows points in the 4 quadrants corresponding to given coordinates x, y.

4.2.1 Compute the distances OP for the points of Fig. 4.2 and determine by trigonometry the corresponding angles θ.

For example:

$$r_{\text{II}} = \sqrt{(-2)^2 + (1)^2} = \sqrt{5} = 2.24,$$

$$\tan \theta_{\text{II}} = \frac{1}{-2} = -0.5, \quad 0.5 \simeq \tan 27°,$$

$$\theta_{\text{II}} = 180° - 27° = 153°.$$

$$(r_{\text{I}} = 2.82,\ \theta_{\text{I}} = 45°;\ r_{\text{III}} = 2.24,\ \theta_{\text{III}} = 207°;$$
$$r_{\text{IV}} = 1.41,\ \theta_{\text{IV}} = -45° = 315°.)$$

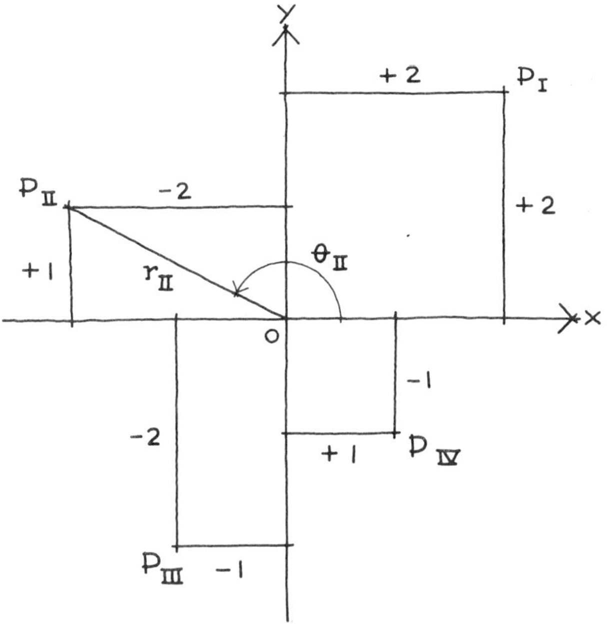

Figure 4.2

By introducing the two coordinates x, y, the concept of point, which is essentially geometrical, has been translated into the concept of a couple of numbers or letters, which is algebraic.

4.3 DISTANCES

Given two points: P_1, of coordinates x_1, y_1, and P_2, of coordinates x_2, y_2 (Fig. 4.3), their distance d is obtained by Pythagoras' theorem:

$$d = \sqrt{(x_2 - x_1)^2 + (y_2 - y_1)^2}. \tag{4.3.1}$$

By trigonometry, the slope m of the line $P_1 P_2$, that is, the tangent of the angle θ between the positive x-axis and the line $P_1 P_2$ is given by:

$$m = \tan\theta = \frac{y_2 - y_1}{x_2 - x_1}. \tag{4.3.2}$$

4.3.1 Compute the distances between P_I and P_{II}, P_I and P_{III}, and P_I and P_{IV} in Fig. 4.2, and the corresponding line slopes. For example, making $P_1 \equiv P_I$,

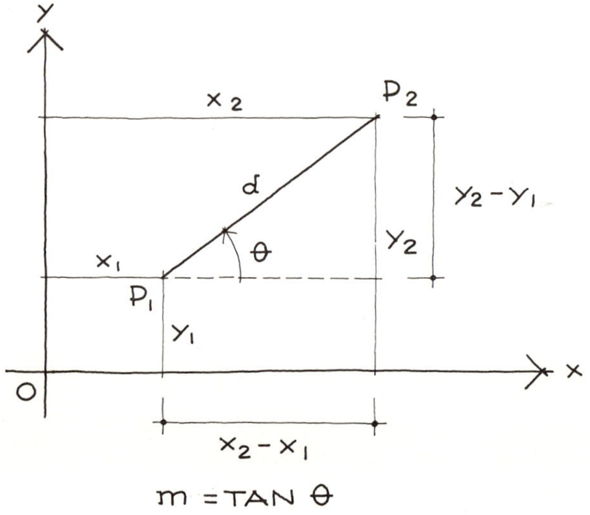

Figure 4.3

$P_2 \equiv P_{\mathrm{III}}$ in (4.3.1):

$$P_{\mathrm{I}}P_{\mathrm{III}} = \sqrt{(-1-2)^2 + (-2-2)^2} = \sqrt{9+16} = \sqrt{25} = 5,$$

$$m_{\mathrm{I,III}} = \frac{-2-2}{-1-2} = \frac{-4}{-3} = \frac{4}{3} = 1.33.$$

$$(P_{\mathrm{I}}P_{\mathrm{II}} = 4.12,\ m_{\mathrm{I,II}} = -0.25;\ P_{\mathrm{I}}P_{\mathrm{IV}} = 3.16,\ m_{\mathrm{I,IV}} = -3.)$$

By (4.3.2), when $y_1 = y_2$, $m = 0$ and the line P_1P_2 has zero slope, i.e., is parallel to the x-axis. When x_1 is almost equal to x_2, $\tan\theta$ is very large (because any number $y_2 - y_1$ different from zero divided by a very small number gives a very large quotient) and when $x_1 = x_2$ we say that $\tan\theta$ is *infinite*. In this case P_1 and P_2 are one on top of the other and P_1P_2 is parallel to the y-axis.

4.4 STRAIGHT LINES

We now wish to give the algebraic equivalent to the geometrical concept of straight line.

A straight line is a line of *constant slope m*. The slope of the line through the two points P_1 and P_2 is given by (4.3.2), but any other two points on this line could be used to determine the value of m. For example, instead of P_2, any other point P of coordinates x, y on the line P_1P_2 of

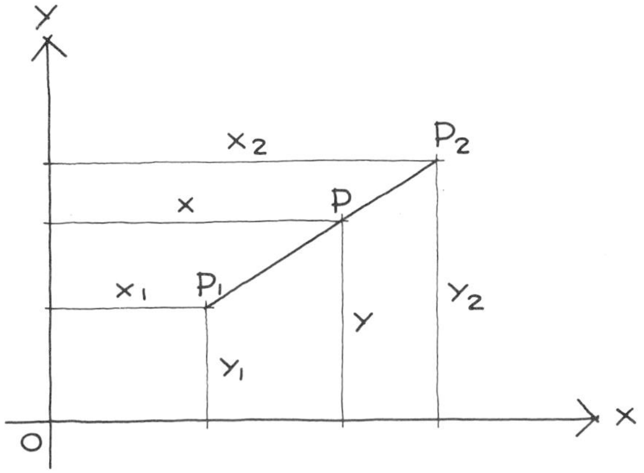

Figure 4.4

Fig. 4.4 could be used together with P_1 to obtain:

$$m = \frac{y - y_1}{x - x_1}. \tag{a}$$

Since the m of (a) and of (4.3.2) are equal:

$$\frac{y - y_1}{x - x_1} = \frac{y_2 - y_1}{x_2 - x_1}. \tag{4.4.1}$$

Because the coordinates of any point P on P_1P_2 satisfy equation (4.4.1), (4.4.1) is called the *equation of the line through* P_1, P_2. For example, from Fig. 4.5 with:

$$x_1 = 1, \quad y_1 = 2; \qquad x_2 = 3, \quad y_2 = 4,$$

(4.3.2) gives

$$m = \frac{4 - 2}{3 - 1} = 1,$$

and the equation of the line through $P_1(1, 2)$ and $P_2(3, 4)$ becomes by (4.4.1):

$$\frac{y - 2}{x - 1} = 1 \qquad \therefore \quad y - 2 = x - 1$$

or:

$$y = x + 1. \tag{b}$$

Equation (b) allows the evaluation of the ordinate y of any point on the line once its abscissa x is known or vice versa. For example, for $x = 2$, $y = 3$; for $y = 1$, $x = 0$. The point B of abscissa $x = 0$ has an ordinate

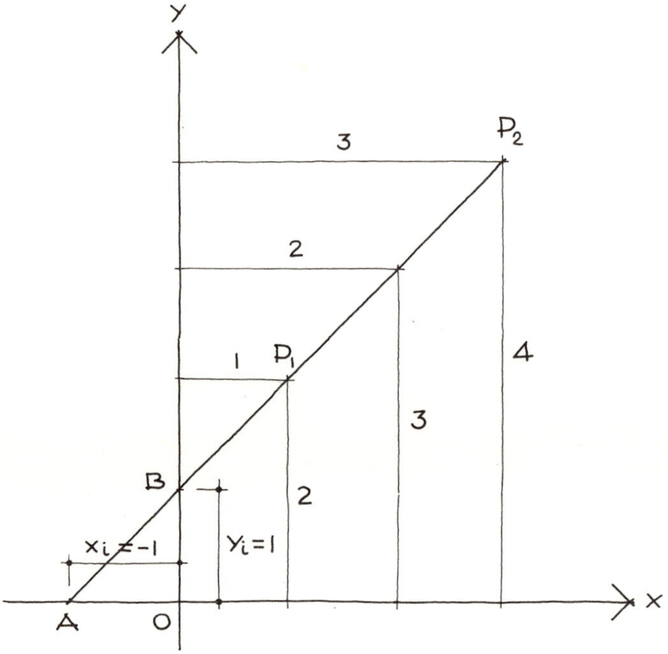

Figure 4.5

y_i called the *y-intercept* of the line P_1P_2. In our example, we obtain from (b) $y_i = 1$. The point A of ordinate $y = 0$ has an abscissa x_i called the *x-intercept* of P_1P_2. In our example, we obtain from (b) $x_i = -1$.

The equation of the line going through the point $P_1(x_1, y_1)$ and having a *given* slope m is obtained from (a), where m is now a given number. Solving for y, we obtain:

$$y - y_1 = m(x - x_1) \qquad \therefore \quad y = mx + (y_1 - mx_1). \qquad (4.4.2)$$

For example, the equation of the line through $P_1(-1, 2)$ with a slope $m = 0.5$ is:

$$y = 0.5x + \{2 - [0.5 \times (-1)]\},$$

or:

$$y = 0.5x + 2.5.$$

The intercepts of this line are:

$$y_i = 2.5; \; x_i = \frac{-2.5}{0.5} = -5.$$

Write the equations of the following lines and compute their intercepts and slopes (when the slope is not given).

4.4.1 Line through $P_1(0, -1)$, $P_2(2, -3)$. $(y = -x - 1, \quad m = -1.)$

4.4.2 Line through $P_1(2, -3)$, $P_2(1, -2)$. $(y = -x - 1, \quad m = -1.)$

4.4.3 Line through $P_1(2, 3)$ with $m = 1$. $(y = x + 1.)$

4.4.4 Horizontal line through $P_1(1, 2)$. $(y = 2.)$

4.4.5 Vertical line through $P_1(-1, -1)$. $(x = -1.)$

Hint: Vertical lines have an infinite m.

Two lines are parallel when they have the same slope. Hence the equation of the line through a point $P_0(x_0, y_0)$ parallel to the line through $P_1(x_1, y_1)$, $P_2(x_2, y_2)$ is obtained by: (a) evaluating the slope m of P_1P_2 by (4.3.2), (b) using m in (4.4.2) with P_0 substituted for P_1:

$$y = mx + (y_0 - mx_0). \tag{4.4.3}$$

For example, given:

$$P_0(1, 4); \qquad P_1(-1, 2), \quad P_2(3, -1),$$

the slope of P_1P_2 is:

$$m = \frac{-1 - 2}{3 - (-1)} = \frac{-3}{4} = -\frac{3}{4},$$

and the line through P_0 parallel to P_1P_2 has the equation:

$$y = -\tfrac{3}{4}x + [4 - (-\tfrac{3}{4}) \times 1] = -\tfrac{3}{4}x + \tfrac{19}{4}.$$

By the third of equations (3.5.1) the slope m_1 of a line at right angles to a line of slope $m = \tan\theta$ is:

$$m_1 = \tan(\theta + 90°) = -\frac{1}{\tan\theta} = -\frac{1}{m}. \tag{4.4.4}$$

Hence, the line through $P_0(x_0, y_0)$ *perpendicular* to a line of slope m is:

$$y = -\frac{1}{m}x + \left[y_0 - \left(-\frac{1}{m}\right)x_0\right] = -\frac{1}{m}x + \left(y_0 + \frac{1}{m}x_0\right). \tag{4.4.5}$$

For example, given:

$$P_0(0, 0), \quad P_1(1, 1), \quad P_2(3, 2)$$

the equation of the line through P_0 perpendicular to the line $P_1 P_2$ is evaluated as follows:

$$m = \frac{2 - 1}{3 - 1} = \frac{1}{2}; \qquad m_1 = -\frac{1}{\frac{1}{2}} = -2;$$

$$y = -2x + (0 - 2 \times 0) = -2x.$$

Write the equations of the following lines.

4.4.6 Parallel to the line $P_1(2, -1)$, $P_2(0, 1)$ through $P_0(3, -7)$.

$$(y = -x - 4.)$$

4.4.7 Perpendicular to the line $y = 3x$ through $P_0(-1, +2)$.

$$(y = -\tfrac{1}{3}x + \tfrac{5}{6}.)$$

4.4.8 Perpendicular to the line $P_1(0, 0)$, $P_2(1, 1)$ through $P_0(0, 0)$.

$$(y = -x.)$$

You might have noticed that all the line equations we have obtained [(4.4.1), (4.4.2), (4.4.3), (4.4.5)] contain both x and y *at the first power*.

The general form of the equation of a straight line also contains x and y at the first power and, hence, is:

$$ay + bx + c = 0, \tag{4.4.6}$$

where a, b and c are constants.

Solving for y, (4.4.6) becomes:

$$y = -\frac{b}{a}x - \frac{c}{a} \tag{4.4.7}$$

and shows by comparison with (4.4.2) that the slope of the line (4.4.6) is:

$$m = -\frac{b}{a}, \tag{4.4.8}$$

while its intercepts are:

$$y_i = -\frac{c}{a}; \qquad x_i = -\frac{c}{b}. \tag{4.4.9}$$

Determine the slope and the intercepts of the following lines.

4.4.9 $3y - x + 2 = 0.$ $(m = \frac{1}{3},\ y_i = -\frac{2}{3},\ x_i = 2.)$

4.4.10 $y + 2x - 2 = 0.$ $(m = -2,\ y_i = 2,\ x_i = 1.)$

4.4.11 $y + x = 0.$ $(m = -1,\ y_i = 0,\ x_i = 0.)$

4.4.12 $x - 3 = 0.$ $(m = \text{infinite},\ x_i = 3.)$

4.5 STRAIGHT LINE APPLICATIONS

As a typical example of the "algebrizing" of geometry by analytic geometry consider the following problem. A cost analysis shows that, on a given site, a 10-story building costs \$1,500,000, a 20-story building \$2,000,000 and a 30-story building \$3,400,000. We would like to evaluate approximately the cost of a 23-story building without going through a lengthy cost analysis. This can be easily done provided we accept a basic simplifying assumption concerning the relationship between the number of stories and cost: that the cost varies as a straight line with the number of stories for buildings with 20 to 30 stories. This assumption is shown in Fig. 4.6, where x is the number of stories and y is the cost in millions of dollars.

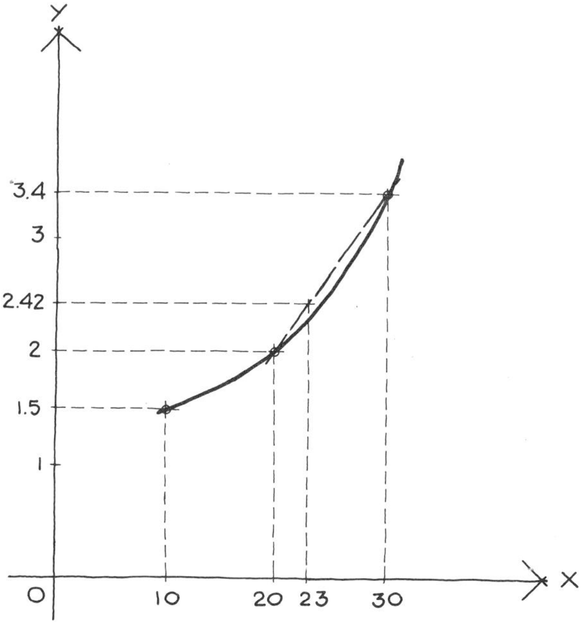

Figure 4.6

By (4.4.1) the straight line $P_1(20, 2)$, $P_2(30, 3.4)$ has the equation:

$$\frac{y-2}{x-20} = \frac{3.4-2}{30-20} = \frac{1.4}{10} = 0.14,$$

from which:

$$y - 2 = 0.14(x - 20)$$

$$\therefore \quad y = 0.14x - 0.8. \tag{a}$$

For $x = 23$, $y = 2.42$, i.e., the cost of a 23-story building is approximately \$2,420,000.

Equation (a) may be used to solve approximately the inverse problem of determining the number of stories in a building of given cost. For example, setting $y = 2.7$ (\$2,700,000) in (a) and solving for x:

$$x = \frac{2.7 + 0.8}{0.14} = 25 \text{ stories.}$$

In general, solving (4.4.1) for y:

$$y = y_1 + (x - x_1)\frac{y_2 - y_1}{x_2 - x_1},$$

and calling x_3 the abscissa at which we want the value y_3 of y:

$$y_3 = y_1 + \frac{x_3 - x_1}{x_2 - x_1}(y_2 - y_1). \tag{4.5.1}$$

Equation (4.5.1) is called a *linear interpolation formula* and is used to obtain an approximate value of y corresponding to a given value of x.

Similarly solving (4.4.1) for x, one obtains an approximate value x_3 of x corresponding to a given value y_3 of y:

$$x_3 = x_1 + \frac{y_3 - y_1}{y_2 - y_1}(x_2 - x_1). \tag{4.5.2}$$

In particular, the value $\bar{x}_3$ of x corresponding to $y_3 = 0$ (Fig. 4.7) is given by:

$$\bar{x}_3 = x_1 - \frac{y_1(x_2 - x_1)}{y_2 - y_1},$$

or bringing x_1 over the same denominator $y_2 - y_1$:

$$\bar{x}_3 = \frac{x_1(y_2 - y_1) - y_1(x_2 - x_1)}{y_2 - y_1}$$

$$= \frac{x_1 y_2 - x_1 y_1 - y_1 x_2 + x_1 y_1}{y_2 - y_1},$$

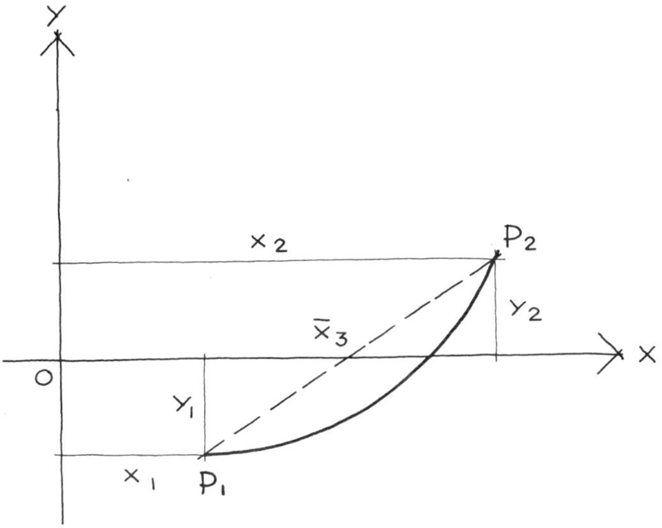

Figure 4.7

so that:

$$\bar{x}_3 = \frac{y_2 x_1 - y_1 x_2}{y_2 - y_1}. \tag{4.5.3}$$

Equation (4.5.3) is most useful in determining "break-even points" and in solving algebraic equations. For example, let us determine the value of x for which the polynomial:

$$y = x^3 + x + 1$$

becomes zero.

The polynomial has the values $y_1 = -1$ at $x_1 = -1$ and $y_2 = +1$ at $x_2 = 0$. Hence, assuming that it has a straight line graph between $x_1 = -1$ and $x_2 = 0$ (Fig. 4.8), it has a value zero at about:

$$\bar{x}_3 = \frac{1 \times (-1) - (-1) \times 0}{1 - (-1)} = \frac{-1}{2} = -\frac{1}{2}.$$

The value of x that makes $y = 0$ is called *a root of the algebraic equation* $y = 0$ (see Section 2.6). Hence, $x = -\frac{1}{2}$ is an approximate root of the cubic equation $x^3 + x + 1 = 0$.

Actually at $x = -\frac{1}{2}$,

$$y = (-\tfrac{1}{2})^3 + (-\tfrac{1}{2}) + 1 = -\tfrac{1}{8} - \tfrac{1}{2} + 1 = \tfrac{3}{8} = 0.375,$$

so that y is positive at $x = -\frac{1}{2}$ and negative at $x = -1$.

If we now assume that y has a straight line graph in the (smaller) interval $x_1 = -1$ (where $y_1 = -1$), $x_2 = -0.5$ (where $y_2 = +0.375$), we can use (4.5.3) again to obtain a more accurate result $\bar{x}_3'$ (Fig. 4.8):

$$\bar{x}_3' = \frac{0.375 \times (-1) - (-1) \times (-0.5)}{0.375 - (-1)} = \frac{-0.375 - 0.5}{1.375} = -0.64.$$

At $\bar{x}_3' = -0.64$,

$$y = (-0.64)^3 + (-0.64) + 1 = 0.098,$$

and with $x_2 = -0.64$, $y_2 = 0.098$, Eq. (4.5.3) gives (Fig. 4.8):

$$\bar{x}_3'' = \frac{0.098 \times (-1) - (-1)(-0.64)}{0.098 - (-1)} = -\frac{0.738}{1.098} = -0.67.$$

At $\bar{x}_3'' = -0.672$, $y = (-0.672)^3 + (-0.672) + 1 = 0.025$, and one more interpolation between $x_1 = -1, y_1 = -1$ and $x_2 = -0.67, y_2 = 0.025$ gives:

$$\bar{x}_3''' = \frac{0.025 \times (-1) - (-1)(-0.67)}{0.025 - (-1)} = -\frac{0.694}{1.025} = -0.68.$$

At $x = -0.68$, $y = 0.0056$, which is small enough to be considered practically zero. Hence, $x = -0.68$ is to 2 significant figures a root of our cubic equation.

Thus, the equation of the straight line allows both interpolation from a table, as was done in the cost problem, and solution of algebraic equations by successive approximations, or the so-called *chords method*, which is illustrated in Fig. 4.8. (In connection with the chords method, you should remember that the quickest way of evaluating a polynomial y of x is by synthetic substitution, as shown in Section 2.5.)

4.5.1 Interpolate y at $x_3 = 2$, if $y_1 = 0$ at $x_1 = 1$ and $y_2 = 6.6$ at $x_2 = 4$. ($y_3 = 2.2$.)

4.5.2 Interpolate y at $x_3 = -0.2$, if $y_1 = -1.6$ at $x_1 = -1$, $y_2 = 3.2$ at $x_2 = 1$. ($y_3 = 0.32$.)

4.5.3 Interpolate x at $y_3 = 60$, if $y_1 = 48$ at $x_1 = 90$, $y_2 = 72$ at $x_2 = 200$. ($x_3 = 145$.)

Determine to 2 figures by successive approximations, using the chords method and synthetic substitution, the root of the following equations in the given intervals.

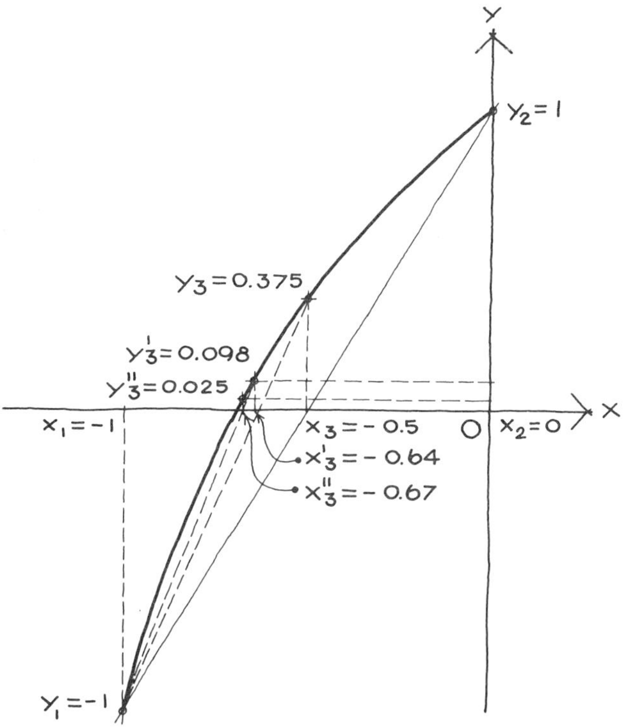

Figure 4.8

4.5.4 $x^3 + 2x^2 + x - 1 = 0$; $x_1 = 0$, $x_2 = 1$. $(x = 0.46.)$

4.5.5 $x^3 + 2x + 1 = 0$; $x_1 = -1$, $x_2 = 0$. $(x = -0.47.)$

4.5.6 $x^4 + x^3 - x^2 + x - 3 = 0$; $x_1 = 1$, $x_2 = 2$. $(x = 1.14.)$

4.5.7 $x^3 - 2x^2 + x - 1 = 0$; $x_1 = 1$, $x_2 = 2$. $(x = 1.76.)$

To illustrate the other aspect of analytic geometry, the "visualization" of algebra, consider a set of two linear equations in the two unknowns x and y:

$$\begin{aligned} 2x + 3y &= 5, \\ -x + 2y &= 1, \end{aligned} \tag{a}$$

and solve them both for y:

$$\begin{aligned} &\text{I.} \qquad y = -\tfrac{2}{3}x + \tfrac{5}{3}, \\ &\text{II.} \qquad y = \tfrac{1}{2}x + \tfrac{1}{2}. \end{aligned} \tag{b}$$

The two equations (b) are the equations of two straight lines (plotted in Fig. 4.9), which intersect at the point (1, 1). The point $x = 1$, $y = 1$ is on both lines, that is, the values $x = 1$, $y = 1$ satisfy both equations (b) and, hence, are the solutions of the set of linear equations (a). Thus, the solution of equations (a) is geometrically equivalent to determining the intersection of two straight lines.

The next section contains additional examples of the visualization of algebra.

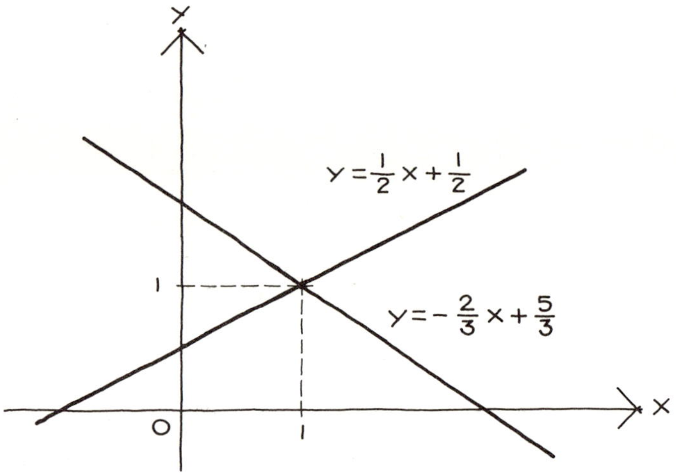

Figure 4.9

4.6 CURVES AND FUNCTIONS

The equation of a straight line is characterized by the fact that the two coordinates x and y appear in it at the first power. An equation in which x and/or y appears at powers different from the first represents always a curve.

For example, the equation:

$$y = 2 + \tfrac{1}{2}x^2 \qquad \text{(a)}$$

is plotted in Fig. 4.10, point by point, by giving to x the values . . . , −3, −2, −1, 0, 1, 2, 3, ...

Similarly, the equation:

$$x^2 + y^2 = 16 \qquad \text{(b)}$$

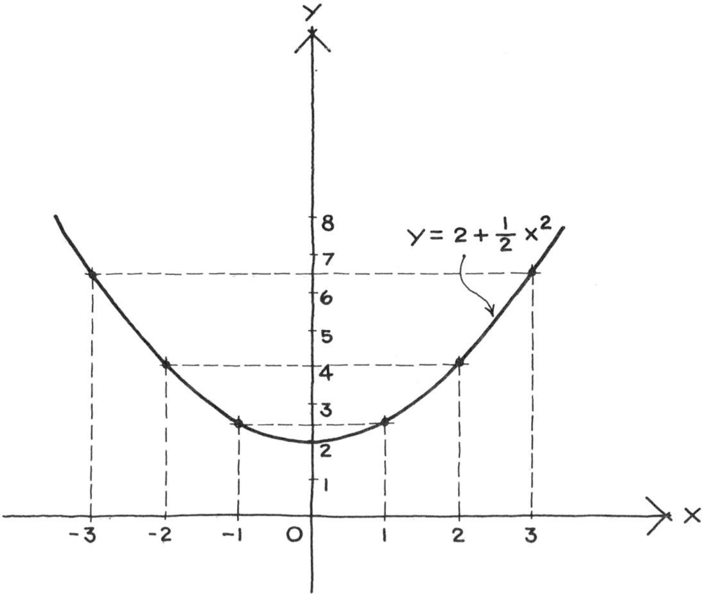

Figure 4.10

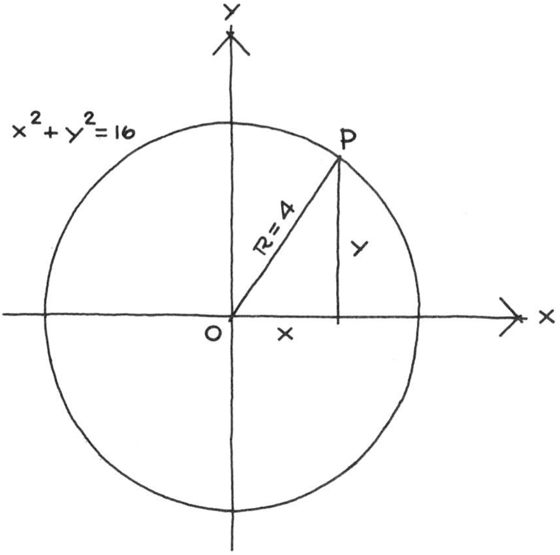

Figure 4.11

is plotted in Fig. 4.11 by solving it first for y:

$$y = \pm\sqrt{16 - x^2}, \tag{b'}$$

and by giving to x the values $-4, -3, -2, -1, 0, 1, 2, 3, 4$.*

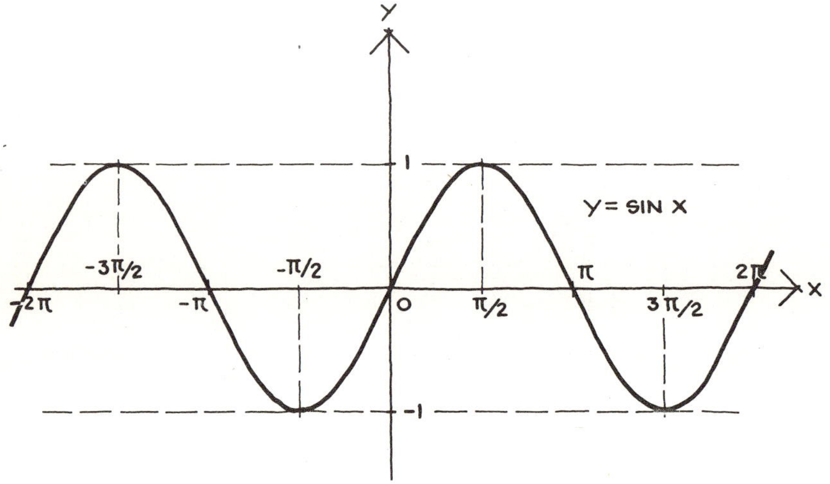

Figure 4.12(a)

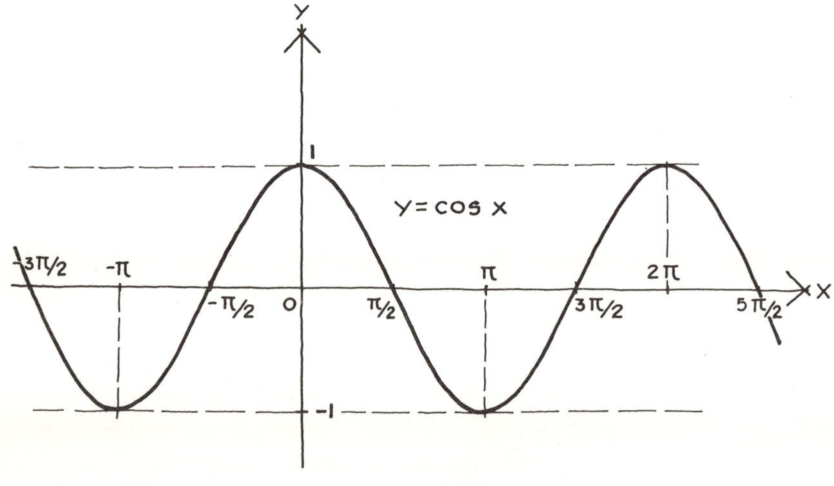

Figure 4.12(b)

*Notice that for $x > 4$ or $x < -4$ the quantity under the square root becomes negative and y is not real. Hence, the curve y does not exist for $x > 4$ or $x < -4$.

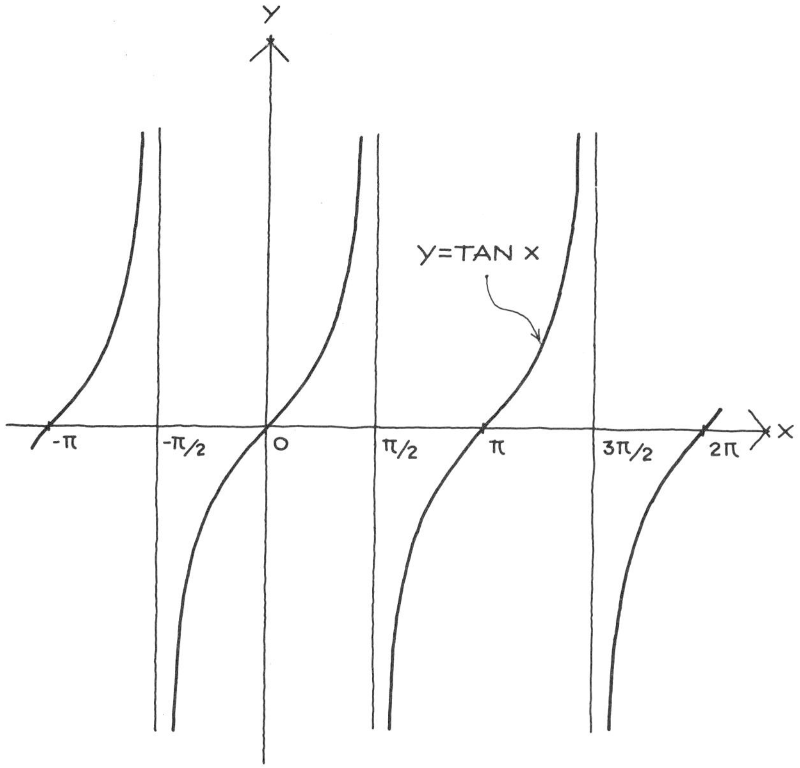

Figure 4.12(c)

The corresponding curve is a circle of center O and radius 4, since $\sqrt{x^2 + y^2}$ is the distance of the point $P(x, y)$ from the origin.

By means of a table of $\sin x$, $\cos x$ and $\tan x$ (Table 2), you may plot the corresponding curves which appear in Figs. 4.12a, 4.12b, 4.12c.

In order to state symbolically that the value of y depends on the value of x we use the notation:

$$y = f(x), \tag{4.6.1}$$

read "y equals f of x" or "y is a function of x."

A function $y = f(x)$ may be graphed, as was done above, on the x, y-plane and, in general, represents a curve.

Plot the following curves between the given values of x.

4.6.1 $y = 3x^3 - 1;\quad x_1 = -3,\ x_2 = 3.$

4.6.2 $y = \sqrt{10 - x^2};\quad x_1 = -\sqrt{10},\ x_2 = \sqrt{10}.$

4.6.3 $y = \dfrac{3x}{4x^2 + 1}$; $x_1 = 0$, $x_2 = 10$.

4.6.4 $x^2 + 3y^2 = 10$; $x_1 = -\sqrt{10}$, $x_2 = \sqrt{10}$.

Hint: Solve for $y = \pm\dfrac{1}{\sqrt{3}}\sqrt{10 - x^2}$.

4.6.5 $y = 2^x$; $x_1 = -2$, $x_2 = 2$.

The graph of a function allows the approximate solution, by the chords method of successive approximations (Section 4.5), of equations which are not represented by polynomials and, in particular, of "break-even" problems.

Assume, for example, that the cost per floor c of a building decreases with the number of floors x according to the law or formula:

$$c = \frac{1}{3 + 0.2x},$$

while the capitalized renting income per floor r grows with x according to the law:

$$r = 0.03x.$$

For what number of floors does the building break even?
If c is to be equal to r:

$$\frac{1}{3 + 0.2x} = 0.03x.* \qquad \text{(c)}$$

Figure 4.13 contains the plots of the curves c and r for x varying between zero and 10. The graphs show that the two curves cross at a point P of abscissa x between 7 and 9 floors. Hence, the break-even point is approximately 8 floors.

If (c) were to be solved with greater accuracy to know whether 7, 8

*This equation is equivalent to the polynomial (quadratic) equation:

$$0.03x(3 + 0.2x) = 1 \text{ or } 0.006x^2 + 0.09x - 1 = 0$$

with roots $x_1 = 7.43$, $x_2 = -22.43$. Equations which cannot be transformed into a polynomial equated to zero are called *transcendental.* For example:

$$\sin x - \tfrac{1}{2}x = 0$$

is a transcendental equation.

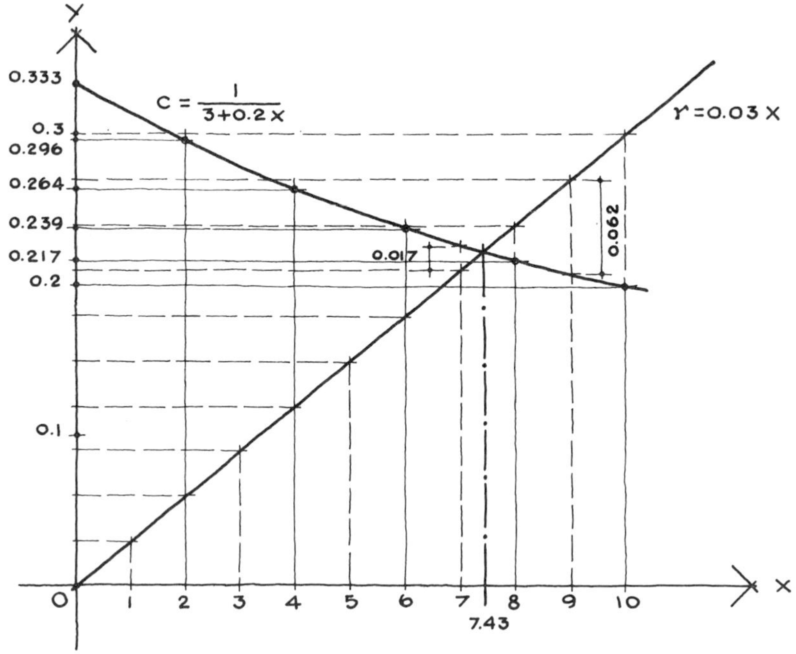

Figure 4.13

or 9 is nearer to the break-even point, we would notice that:

$$\text{at } x = 7: \quad c = 0.227, \quad r = 0.210$$
$$\text{at } x = 9: \quad c = 0.208, \quad r = 0.270$$

so that $c - r = +0.017$ is positive at $x = 7$ and $c - r = -0.062$ is negative at $x = 9$. To find where $c - r = 0$ we use linear interpolation [Eq. (4.5.3)] with $c - r = y$ and:

$$x_1 = 7, \quad y_1 = 0.017; \qquad x_2 = 9, \quad y_2 = -0.062,$$

obtaining:

$$\bar{x}_3 = \frac{-0.062 \times 7 - 0.017 \times 9}{-0.062 - 0.017} = \frac{-0.587}{-0.079} = 7.43.$$

Since 7.43 is nearer to 7 than to 8, 7 floors is the break-even point.

Find to 2 figures the root of the following equations in the given interval x_1, x_2 by the chords method of successive approximations. *Hint:* Graph the functions in the given interval first.

4.6.6 $\frac{1}{x+2} = \sqrt{1-x^2}$; $x_1 = 0$, $x_2 = 1$. ($x = 0.94$.)

4.6.7 $(x+3)^2 = -\frac{1}{x}$; $x_1 = -2$, $x_2 = -3$. ($x = -2.35$.)

4.6.8 $2 \sin x = 3 \cos x$; $x_1 = 50°$, $x_2 = 60°$. ($x = 56°$.)

4.6.9 $\tan x = x$; $x_1 = 180°$, $x_2 = 270°$. ($x = 257° = 4.49$ rad.)

4.6.10 $\sin^2 x = \cos x$; $x_1 = 45°$, $x_2 = 55°$. ($x = 52°$.)

5

DIFFERENTIAL CALCULUS

5.1 INTRODUCTION

Contrary to common belief the calculus is conceptually and practically easier to learn than algebra. We use it, without even noticing it, in our daily life. For example, the statement: "My car is doing 55 miles an hour" is a statement that involves the calculus. So does the statement: "The area of that free form swimming pool is 500 square feet."

The calculus deals mostly with quantities that change value, while algebra deals with constant quantities. Since most of the quantities encountered in life change value, the calculus is one of the basic tools of applied mathematics.

As examples of quantities that change value you may want to think of the change in elevation of a road with the distance from a town, of the change in temperature of the air with the time of day, or of the cost of a building with the number of its floors. You will find the calculus a simple tool to answer questions like: "For what number of floors will a building's cost per floor be a minimum?"; "What is the shallowest beam I can use to support a given load over a span of given length?"; "What is the largest area I can fence with a given length of fence?"; "What is the sag of a cable-supported roof which will result in the most economical cable of given span?"

A *careful* reading of the sections of this and the following chapter will convince you of the truth of the statements made above about the usefulness and simplicity of the calculus.

5.2 THE SLOPE OF A CURVE

The slope of a straight line was seen in Section 4.4 to be the tangent of the angle θ between the positive x-axis and the straight line. In practice, we often refer to the slope at a point of a road, a river or a mountain, but the profiles we consider in these cases are not straight lines, they are curves. Therefore, we need to define and evaluate the slope at a point of a curve. We shall do this "by successive approximations" starting with the slope of a chord of the curve.

Let us consider the curve of Fig. 5.1 and try to define its "slope" at the point P_1 of coordinates x_1, y_1. To this purpose we choose on the curve

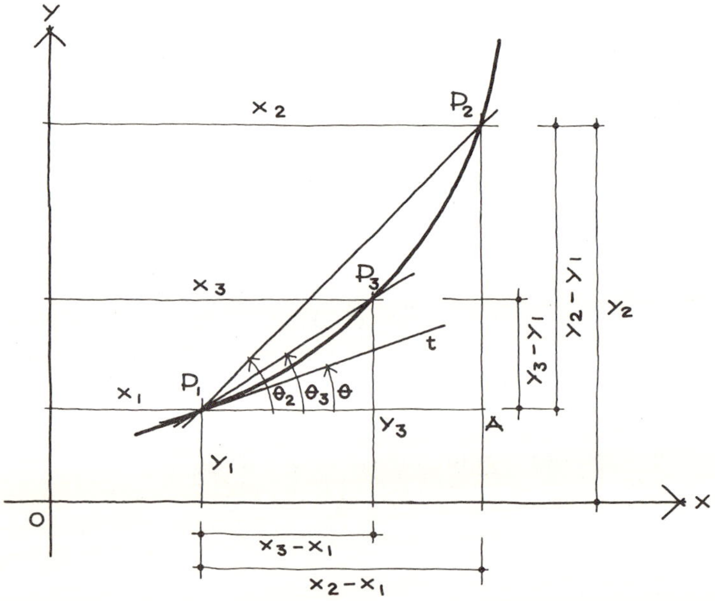

Figure 5.1

a point P_2 (x_2, y_2) and evaluate first the slope m_2 of the straight line or *chord* P_1P_2. By the definition of $\tan \theta_2$:

$$m_2 \equiv \tan \theta_2 = \frac{AP_2}{P_1A} = \frac{y_2 - y_1}{x_2 - x_1}, \qquad (5.2.1)$$

where m_2 is the *average slope* of the curve between P_1 and P_2. In Fig. 5.1 it is larger than what we intuitively think of as the slope of the curve at P_1.

Let us now choose a new point P_3 (x_3, y_3) *nearer* to P_1 than P_2, that is, with $x_3 - x_1$ smaller than $x_2 - x_1$. The slope of the chord P_1P_3 is:

$$m_3 = \tan \theta_3 = \frac{y_3 - y_1}{x_3 - x_1}. \qquad (5.2.1')$$

The slope m_3 is nearer to the slope of the curve at P_1 than m_2. It now becomes apparent that to obtain an excellent approximation of the slope at P_1 we should take a point P (x, y) very near P_1, i.e., such that $x - x_1$ be extremely small. In this case $y - y_1$ would also become extremely small and the ratio $(y - y_1)/(x - x_1)$ would be the ratio of two very small numbers. If we could only get P on top of P_1, the evaluation of the slope would be theoretically "perfect," but it could not really be obtained because both $y - y_1$ and $x - x_1$ would become zero and the expression $m = 0/0$ has no meaning. (To realize why $0/0$ has no meaning just think that zero divided by any small number $x - x_1$ is zero, while any small number $y - y_1$ divided by zero is infinitely large. Thus, $m = 0/0$ may be anywhere between zero and infinity.) Since geometrically as x approaches x_1 the chord approaches the *line t tangent to the curve at* P_1 (Fig. 5.1) and since this line has a *specific* slope $m = \tan \theta$, we must establish a procedure for obtaining such slope, which we consider and call the *slope of the curve at* P_1. This procedure is called *differentiation by the delta method.*

Let us indicate the difference between any value of y and the value $y = y_1$ in Fig. 5.2 by the shorthand symbol:

$$\Delta y_1 = y - y_1, \qquad (5.2.2)$$

where the Greek letter Δ (read "delta") preceding y_1 indicates that y_1 must be subtracted from y. In other words, Δy is *not* the algebraic symbol for the product of Δ by y; it is a calculus symbol that indicates the difference $y - y_1$. The shorthand symbol:

$$\Delta x_1 = x - x_1 \qquad (5.2.3)$$

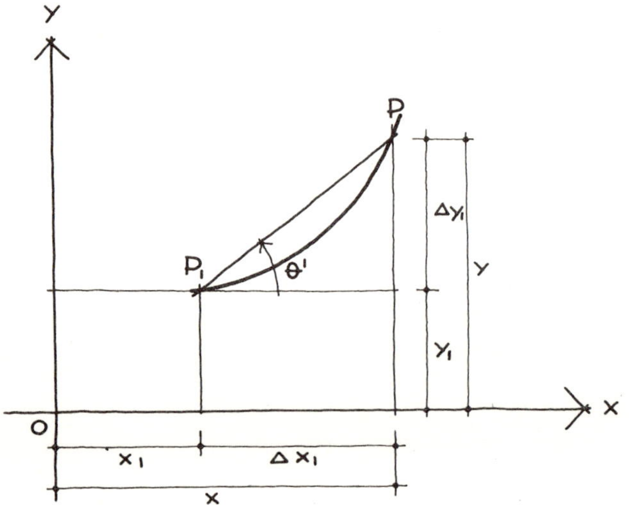

Figure 5.2

means, similarly, the difference between any value of x and the value $x = x_1$. The slope of the chord P_1P (Fig. 5.2) is then given by:

$$m' = \tan \theta' = \frac{\Delta y_1}{\Delta x_1}. \tag{5.2.4}$$

To obtain the slope m of the tangent line t at P_1 we now make Δx_1 smaller and smaller, evaluate the corresponding smaller and smaller values of Δy_1 and plot the ratios $\Delta y_1/\Delta x_1$ of (5.2.4). Since we cannot *make* Δx_1 *equal to zero*, we plot the ratios $\Delta y_1/\Delta x_1$ versus Δx_1 and see whether there is a trend to these ratios. The value the ratios approach (if there is a trend) is the value of the slope m of the tangent line t at P_1, i.e., the slope of the curve at P_1.

For example, consider the curve:

$$y = x^2 \tag{a}$$

of Fig. 5.3 and take P_1 to be the point $x_1 = 1, y_1 = 1$. In Table 5.1 the points P have successively $x = 2, 1.5, 1.25, 1.12, 1.06, 1.03, 1.01, 1.001$ and, correspondingly, by (a), $y = (2)^2 = 4$, $(1.5)^2 = 2.25$, $(1.25)^2 = 1.563, \ldots$. The corresponding values of $\Delta x_1 = x - x_1$ and $\Delta y_1 = y - y_1$ are shown in columns 2 and 4, and the ratios $\Delta y_1/\Delta x_1$ in column 5 of Table 5.1.

Inspection of column 5 shows clearly that as x approaches $x_1 = 1$, i.e., as Δx_1 approaches zero, the trend of the ratios $\Delta y_1/\Delta x_1$ is towards the value 2. Hence, the slope of the curve $y = x^2$ at $x = 1$ equals 2, as can be checked graphically by the slope of the tangent line t_1 in Fig. 5.3.

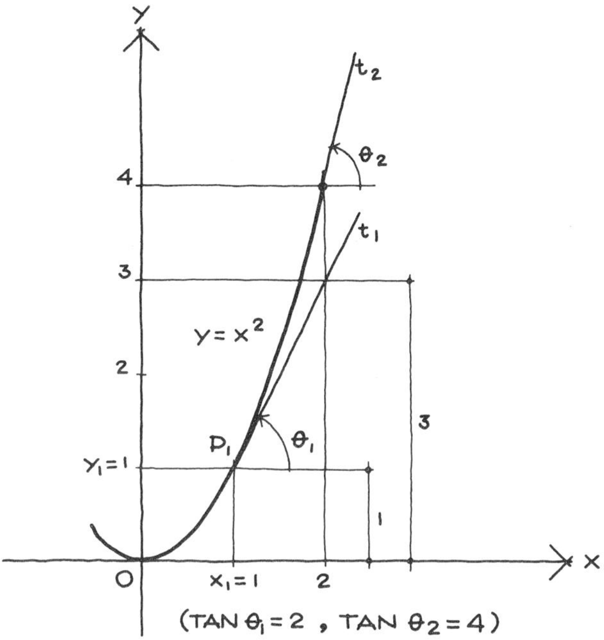

Figure 5.3

Table 5.1

(1) x	(2) $\Delta x = x - x_1$	(3) $y = x^2$	(4) $\Delta y = y - y_1$	(5) $\Delta y_1/\Delta x_1$
2.00	1.00	4.00	3.00	3.00
1.50	0.50	2.25	1.25	2.5
1.25	0.25	1.563	0.563	2.25
1.12	0.12	1.254	0.254	2.125
1.06	0.06	1.123	0.123	2.050
1.03	0.03	1.061	0.061	2.033
1.01	0.01	1.0201	0.0201	2.010
1.001	0.001	1.002001	0.002001	2.001

5.3 THE DERIVATIVE

The process of differentiation, carried out by arithmetic in Table 5.1, may be carried out more efficiently by algebra so as to obtain the slope of the curve $y = x^2$ at *any* point $P_1(x_1, y_1)$.

From (5.2.3) and (5.2.2), the coordinates x, y of P may be written as:

$$x = x_1 + \Delta x_1, \qquad y = y_1 + \Delta y_1. \tag{5.3.1}$$

But with $y = x^2$ and $x = x_1 + \Delta x_1$:

$$y = (x_1 + \Delta x_1)^2 = x_1^2 + 2x_1\,\Delta x_1 + (\Delta x_1)^2$$

[see (2.4.1)] or, since $y_1 = x_1^2$:

$$y = y_1 + 2x_1\,\Delta x_1 + (\Delta x_1)^2,$$

from which:

$$\Delta y_1 \equiv y - y_1 = 2x_1\,\Delta x_1 + (\Delta x_1)^2.$$

Dividing both sides of this equation by Δx_1 we obtain:

$$\frac{\Delta y_1}{\Delta x_1} = 2x_1 + \Delta x_1. \tag{a}$$

The process of making Δx_1 smaller and smaller, that is, of having Δx_1 *approach zero* is indicated in the calculus by the shorthand symbol:

$$\lim_{\Delta x_1 \to 0},$$

(read "limit as Δx_1 approaches zero"). By means of this symbol, the value of the slope m at $x = x_1$ may be written as:

$$m = \lim_{\Delta x_1 \to 0} \frac{\Delta y_1}{\Delta x_1} = \lim_{\Delta x_1 \to 0} (2x_1 + \Delta x_1). \tag{5.3.2}$$

But as Δx_1 approaches zero, $2x_1 + \Delta x_1$ approaches $2x_1$ (since Δx_1 vanishes, while x_1 stays put); hence:

$$m = 2x_1. \tag{5.3.3}$$

By (5.3.3), $m = 2$ at $x_1 = 1$ (as was shown in Table 5.1), $m = 4$ at $x_1 = 2$, $m = 6$ at $x_1 = 3$, etc., as you may check by prolonging the graph of $y = x^2$. In general then, the slope of $y = x^2$ at any point x equals $2x$.

The process of differentiation leads to the derivation of a new function related to $y = x^2$, the function $y = 2x$, which is called the *derivative of* y and is indicated by a variety of symbols, but mostly by dy/dx (not to be interpreted as an algebraic fraction but as a single symbol d/dx applied to y).

It was noticed at the end of Section 4.6 that when a quantity y has values depending on the values of another quantity x (as the ordinates y of a curve depend on their abscissas x) it is convenient to indicate the dependence of y on x by the symbol $y = f(x)$ [see (4.6.1)]. In this symbol $f(x)$ stands for the operations to be performed on x to obtain y. For example: if $y = x^2$, $f(x)$ stands for "square the value of x"; if $y = x^3 + 2$, $f(x)$ stands for "cube x and add 2"; if $y = \sin x$, $f(x)$ stands for "look for the value of the sine of x in a table of sines."

The process of differentiation of a function $y = f(x)$ at $x = x_1$, symbolized by

$$\left.\frac{dy}{dx}\right]_{x=x_1},$$

consists then in performing the following 5 steps of the *delta method:*

$$\begin{aligned}
&\text{(a)}\ y_1 = f(x_1),\\
&\text{(b)}\ y = f(x_1 + \Delta x_1),\\
&\text{(c)}\ \Delta y_1 \equiv y - y_1 = f(x_1 + \Delta x_1) - f(x_1),\\
&\text{(d)}\ \frac{\Delta y_1}{\Delta x_1} = \frac{f(x_1 + \Delta x_1) - f(x_1)}{\Delta x_1},\\
&\text{(e)}\ \left.\frac{dy}{dx}\right]_{x=x_1} = \lim_{\Delta x_1 \to 0} \frac{\Delta y_1}{\Delta x_1}.
\end{aligned} \tag{5.3.4}$$

For example, given:

$$y = x^3,$$

$$\begin{aligned}
&\text{(a)}\ y_1 = x_1^3,\\
&\text{(b)}\ y = (x_1 + \Delta x_1)^3 = x_1^3 + 3x_1^2\,\Delta x_1 + 3x_1(\Delta x_1)^2 + (\Delta x_1)^3 \quad \text{[see (2.4.2)]},\\
&\text{(c)}\ \Delta y_1 \equiv y - y_1 = 3x_1^2\,\Delta x_1 + 3x_1(\Delta x_1)^2 + (\Delta x_1)^3,\\
&\text{(d)}\ \frac{\Delta y_1}{\Delta x_1} = 3x_1^2 + 3x_1\,\Delta x_1 + (\Delta x_1)^2,\\
&\text{(e)}\ \left.\frac{dy}{dx}\right]_{x-x_1} = 3x_1^2,
\end{aligned}$$

since, as Δx_1 approaches zero, both $3x_1\,\Delta x_1$ and $(\Delta x_1)^2$ vanish. [Actually, $(\Delta x_1)^2$ vanishes faster than $3x_1\,\Delta x_1$ for all values of x_1 except $x_1 = 0$.]

We can now obtain the derivative of any power of x by means of Newton's binomial expansion (2.4.4). Given:

$$y = x^n,$$

(a) $y_1 = x_1^n,$

(b) $y = (x_1 + \Delta x_1)^n = x_1^n + nx_1^{n-1}\,\Delta x_1 + \frac{n(n-1)}{2}x_1^{n-2}\,(\Delta x_1)^2 + \frac{n(n-1)(n-2)}{6}x_1^{n-3}\,(\Delta x_1)^3 + \ldots,$

(c) $\Delta y_1 = nx_1^{n-1}\,\Delta x_1 + \frac{n(n-1)}{2}x_1^{n-2}\,(\Delta x_1)^2 + \frac{n(n-1)(n-2)}{6}x_1^{n-3}\,(\Delta x_1)^3 + \ldots,$

(d) $\frac{\Delta y_1}{\Delta x_1} = nx_1^{n-1} + \frac{n(n-1)}{2}x_1^{n-2}\,\Delta x_1 + \frac{n(n-1)(n-2)}{6}x_1^{n-3}\,(\Delta x_1)^2 + \ldots,$

(e) $\left.\frac{dy}{dx}\right]_{x=x_1} = nx_1^{n-1}.$ (5.3.5)

Equation (5.3.5) holds for any value of n, since (2.4.4) holds for any value of n. For example:

$$y = \frac{1}{x} = x^{-1}, \qquad \frac{dy}{dx} = (-1)x^{-1-1} = -x^{-2} = -\frac{1}{x^2}.$$

$$y = \sqrt{x} = x^{1/2}, \qquad \frac{dy}{dx} = \frac{1}{2}x^{1/2-1} = \frac{1}{2}x^{-1/2} = \frac{1}{2}\frac{1}{x^{1/2}} = \frac{1}{2\sqrt{x}}.$$

$$y = \frac{1}{\sqrt[3]{x}} = \frac{1}{x^{1/3}} = x^{-1/3}, \qquad \frac{dy}{dx} = -\frac{1}{3}x^{-1/3-1} = -\frac{1}{3}x^{-4/3} = -\frac{1}{3}\frac{1}{x^{4/3}} = -\frac{1}{3}\frac{1}{\sqrt[3]{x^4}}.$$

We observe, finally, that if a function $y = f(x)$ is multiplied by a constant c, its derivative is multiplied by the same constant, since by the delta process:

(a) $cy_1 = cf(x_1)$

(b) $cy = cf(x_1 + \Delta x_1)$

(c) $\Delta c y_1 = c f(x_1 + \Delta x_1) - c f(x_1) = c[f(x_1 + \Delta x_1) - f(x_1)]$

(d) $\dfrac{\Delta c y_1}{\Delta x_1} = c\dfrac{f(x_1 + \Delta x_1) - f(x_1)}{\Delta x_1}$

(e) $$\left.\frac{d(cy)}{dx}\right]_{x=x_1} = \lim_{\Delta x_1 \to 0} c\frac{\Delta y_1}{\Delta x_1} = c \lim_{\Delta x_1 \to 0} \frac{\Delta y_1}{\Delta x_1} = c\left.\frac{dy}{dx}\right]_{x=x_1} \tag{5.3.6}$$

For example:

$$y = 3x^2; \qquad \frac{dy}{dx} = 3(2x) = 6x.$$

$$y = c\sqrt{x}; \qquad \frac{dy}{dx} = c\left(\frac{1}{2}x^{-1/2}\right) = \frac{c}{2\sqrt{x}}.$$

$$y = \text{a constant } c = c \times 1 = c \times x^0; \quad \frac{dy}{dx} = c(0 \times x^{0-1})$$
$$= 0\frac{c}{x} = 0. \tag{5.3.7}$$

This last result shows that *the derivative of a constant is zero.* This is obvious, since the straight line $y = c$ is parallel to the x-axis, i.e., is horizontal and has zero slope.

It may also be proved by the delta method that *the derivative of the sum of two or more functions is the sum of the derivatives of the functions.* For example, if $y_1 = x^2$, $y_2 = 2/x$ and $y = y_1 + y_2$:

$$y = x^2 + \frac{2}{x} = x^2 + 2x^{-1}.$$

$$\frac{dy}{dx} = \frac{dx^2}{dx} + 2\frac{dx^{-1}}{dx} = 2x + 2(-x^{-2}) = 2x - \frac{2}{x^2}.$$

You may now evaluate the derivatives of the following functions at the given x_1, i.e., the slope of the corresponding curves at the given point (check a few results on the graphs of the functions).

5.3.1 $y = 3x^2 + 2x + 3, \quad x_1 = 2. \qquad \left(\left.\dfrac{dy}{dx}\right]_{x=2} = 14.\right)$

5.3.2 $y = 6x^3 + \dfrac{1}{x}, \quad x_1 = -1. \qquad \left(\left.\dfrac{dy}{dx}\right]_{x=-1} = 17.\right)$

5.3.3 $y = 4\sqrt[3]{x}, \quad x_1 = 8. \qquad \left(\left.\dfrac{dy}{dx}\right]_{x=8} = \dfrac{1}{3}.\right)$

5.3.4 $y = \frac{3}{\sqrt[3]{x}} + 2\sqrt{x}$, $x_1 = 1$. $\left(\frac{dy}{dx}\right]_{x=1} = 0.\Big)$

5.3.5 $y = x^2 - \frac{1}{x^2}$, $x_1 = 2$. $\left(\frac{dy}{dx}\right]_{x=2} = 4.25.\Big)$

5.4 THE DERIVATIVES OF $\sin x$, $\cos x$, $\tan x$

The slopes of the curves of Fig. 4.12, i.e., the derivatives of $\sin x$, $\cos x$ and $\tan x$ may be obtained by the delta method using the formulas for $\sin(\alpha + \beta)$ and $\cos(\alpha + \beta)$ obtained in Section 3.4.

Let in (3.4.8):

$$\alpha = x_1, \qquad \beta = \Delta x_1,$$

so that:

$$\sin(x_1 + \Delta x_1) = \sin x_1 \cos \Delta x_1 + \cos x_1 \sin \Delta x_1, \tag{a}$$

and apply the delta method:

(a) $y_1 = \sin x_1$,

(b) $y = \sin(x_1 + \Delta x_1) = \sin x_1 \cos \Delta x_1 + \cos x_1 \sin \Delta x_1$,

(c) $\Delta y_1 = \sin x_1 \cos \Delta x_1 + \cos x_1 \sin \Delta x_1 - \sin x_1$,

(d) $\dfrac{\Delta y_1}{\Delta x_1} = \dfrac{\sin x_1 \cos \Delta x_1 + \cos x_1 \sin \Delta x_1 - \sin x_1}{\Delta x_1}$.

As Δx_1 approaches zero, by equations (3.7.1) for the evaluation of the trigonometric functions when $x \ll 1$: $\sin \Delta x_1 = \Delta x_1$, $\cos \Delta x_1 = 1$. Hence:

(e) $$\left.\frac{d \sin x}{dx}\right]_{x=x_1} = \lim_{\Delta x_1 \to 0} \frac{\Delta y_1}{\Delta x_1} = \frac{\cancel{\sin x_1} \times 1 + \cos x_1 \times \Delta x_1 - \cancel{\sin x_1}}{\Delta x_1}$$
$$= \cos x_1 \frac{\Delta x_1}{\Delta x_1} = \cos x_1.$$

Thus:

$$\frac{d \sin x}{dx} = \cos x. \tag{5.4.1}$$

Similarly, by (3.4.9) with $\alpha = x_1$, $\beta = \Delta x_1$:

(a) $y_1 = \cos x_1$,

(b) $y = \cos(x_1 + \Delta x_1) = \cos x_1 \cos \Delta x_1 - \sin x_1 \sin \Delta x_1,$

(c) $\Delta y_1 = \cos x_1 \cos \Delta x_1 - \sin x_1 \sin \Delta x_1 - \cos x_1,$

(d) $\dfrac{\Delta y_1}{\Delta x_1} = \dfrac{\cos x_1 \cos \Delta x_1 - \sin x_1 \sin \Delta x_1 - \cos x_1}{\Delta x_1},$

and by (3.7.1):

(e) $$\frac{d \cos x}{dx}\bigg]_{x=x_1} = \lim_{\Delta x_1 \to 0} \frac{\Delta y_1}{\Delta x_1} = \frac{\cancel{\cos x_1} \times 1 - \sin x_1 \times \Delta x_1 - \cancel{\cos x_1}}{\Delta x_1}$$
$$= -\sin x_1 \frac{\Delta x_1}{\Delta x_1} = -\sin x_1.$$

Thus:

$$\frac{d \cos x}{dx} = -\sin x. \tag{5.4.2}$$

By (3.4.10) with $\alpha = x_1$ and $\beta = \Delta x_1$ we obtain:

(a) $y_1 = \tan x_1$

(b) $y = \tan(x_1 + \Delta x_1) = \dfrac{\tan x_1 + \tan \Delta x_1}{1 - \tan x_1 \tan \Delta x_1}$

(c) $\Delta y_1 = \dfrac{\tan x_1 + \tan \Delta x_1}{1 - \tan x_1 \tan \Delta x_1} - \tan x_1$

(d) $\dfrac{\Delta y_1}{\Delta x_1} = \dfrac{1}{\Delta x_1}\left(\dfrac{\tan x_1 + \tan \Delta x_1}{1 - \tan x_1 \tan \Delta x_1} - \tan x_1\right)$

or, bringing $\tan x_1$ over the same denominator $1 - \tan x_1 \tan \Delta x_1$:

$$\frac{\Delta y_1}{\Delta x_1} = \frac{\tan x_1 + \tan \Delta x_1 - \tan x_1(1 - \tan x_1 \tan \Delta x_1)}{\Delta x_1(1 - \tan x_1 \tan \Delta x_1)}$$
$$= \frac{\cancel{\tan x_1} + \tan \Delta x_1 - \cancel{\tan x_1} + \tan^2 x_1 \tan \Delta x_1}{\Delta x_1(1 - \tan x_1 \tan \Delta x_1)}.$$

By (3.7.1) as Δx_1 approaches zero, $\tan \Delta x_1 = \Delta x_1$; hence:

(e) $$\frac{d \tan x}{dx}\bigg]_{x=x_1} = \lim_{\Delta x_1 \to 0} \frac{\Delta y_1}{\Delta x_1} = \frac{\Delta x_1 + \tan^2 x_1 \Delta x_1}{\Delta x_1(1 - \tan x_1 \Delta x_1)} = \frac{1 + \tan^2 x_1}{1 - \tan x_1 \Delta x_1}.$$

In the denominator as $\Delta x_1 \to 0$, $\tan x_1 \Delta x_1$ vanishes and:

$$\frac{d \tan x}{dx}\bigg]_{x=x_1} = 1 + \tan^2 x_1 = 1 + \frac{\sin^2 x_1}{\cos^2 x_1} = \frac{\cos^2 x_1 + \sin^2 x_1}{\cos^2 x_1}.$$

But by (3.4.1) $\sin^2 x + \cos^2 x = 1$, so that:

$$\left.\frac{d \tan x}{dx}\right]_{x=x_1} = \frac{1}{\cos^2 x_1}. \tag{5.4.3}$$

(Notice that although $\tan x$ is the *ratio* of $\sin x$ divided by $\cos x$, the derivative of $\tan x$ is *not the ratio of the derivatives of* $\sin x$ *and of* $\cos x$. You will see in Section 5.7 how to evaluate directly the derivatives of products and ratios of two functions.)

Check the following derivatives at the given points. *Note:* In the calculus the measure x of an angle is *always* in radians.

5.4.1 $\left.\frac{d \sin x}{dx}\right]_{x=0.2} = 0.98.$

5.4.2 $\left.\frac{d \cos x}{dx}\right]_{x=0.7} = -0.64.$

5.4.3 $\left.\frac{d \tan x}{dx}\right]_{x=1.2} = 7.8.$

5.4.4 $\left.\frac{d \tan x}{dx}\right]_{x=\pi} = 1.$

5.4.5 $\left.\frac{d \cos x}{dx}\right]_{x=\pi} = 0.$

5.4.6 $\left.\frac{d \sin x}{dx}\right]_{x=\pi/2} = 0.$

5.5 APPLICATIONS OF THE DERIVATIVE

Whenever a *smooth* curve reaches its highest or lowest point within a given interval (Fig. 5.4), its tangent is horizontal at such point, i.e., the slope of the curve is zero there. (When the curve reaches its highest or lowest point at the boundary of the interval (Fig. 5.5a) or is not smooth but jagged (Figs. 5.5b, c), the slope is not necessarily zero there.) This simple observation explains why problems, in which a quantity is either maximized or minimized (optimal problems, see Section 5.1), are solved by differentiation. Points like A and B in Fig. 5.4 are called *maximum* and *minimum points* of y. Translating the geometrical property of the slope of the curve y at A and B into the shorthand of the calculus, we can say that: *At points where a smooth function* $y = f(x)$ *is maximum or minimum within a given interval, the derivative of* y *is zero.*

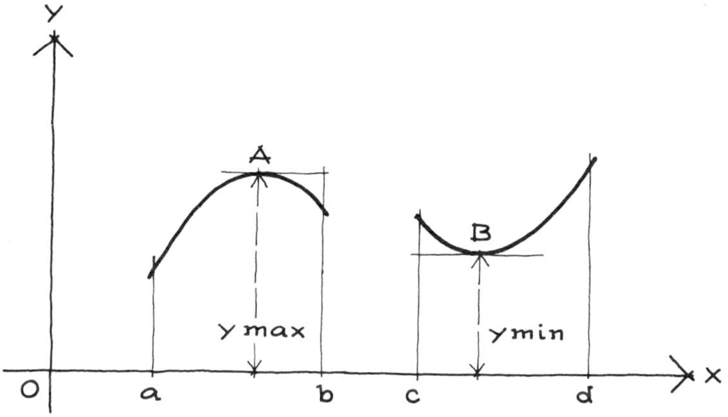

Figure 5.4

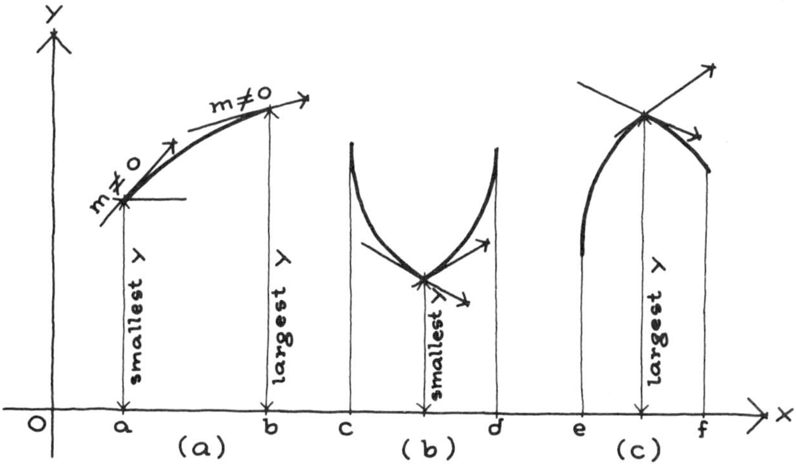

Figure 5.5

You should realize that this calculus statement says, simply: "If you are going up, before you start going down you must stay level for an instant" or, "If you are going down, before you start going up you must stay level for an instant."

Let us use the above statement to solve an optimal problem. You wish to enclose a *rectangular* area (Fig. 5.6) with 100 ft of fence. What is the largest rectangular area you can fence in? Call x and y the sides of the rectangle. Its perimeter equals $2x + 2y$ and must equal the length of your fence. Hence:

$$2x + 2y = 100,$$

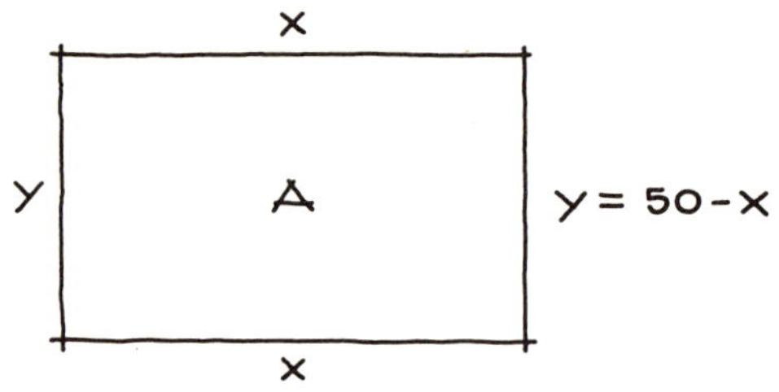

Figure 5.6

from which:

$$2y = 100 - 2x, \quad y = \tfrac{1}{2}(100 - 2x) = 50 - x. \tag{a}$$

The area of the rectangle equals the product of its sides:

$$A = xy = x(50 - x) = 50x - x^2.$$

A is a function of x and is plotted in Fig. 5.7. We wish to find out for what value of x the function A is maximum. Since at such value of x the derivative of A must be zero, evaluating dA/dx by (5.3.5) and setting it equal to zero we

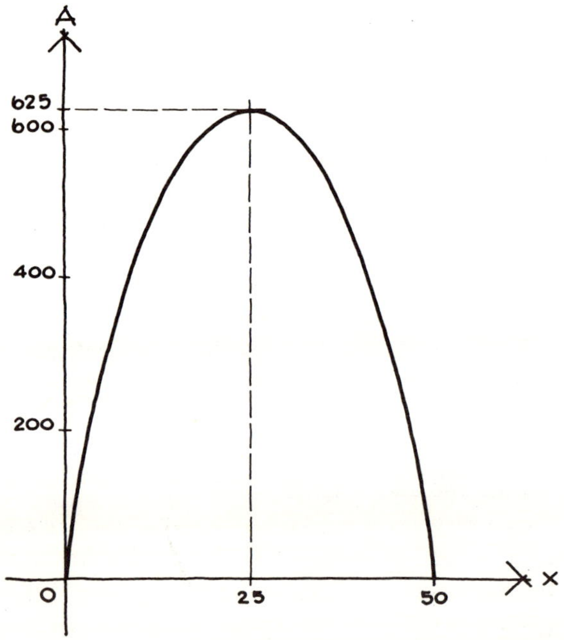

Figure 5.7

obtain:

$$\frac{dA}{dx} = 50 - 2x = 0. \tag{b}$$

(b) is the condition satisfied by the value of x which makes A maximum and, hence, solves the problem. It is an algebraic equation in x, whose only root is $x = 25$ ft. With $x = 25$ ft (a) gives:

$$y = 50 - x = 50 - 25 = 25 \text{ ft}.$$

The maximum rectangular fenced area is the area of a square 25 ft $\times$ 25 ft $= 625$ sq ft. (The maximum area of *any* shape fenced by a perimeter of 100 ft is a circle of radius r given by $2\pi r = 100$, i.e., with $r = 100/2\pi = 16$ ft and an area $A = \pi r^2 = 3.14\,(16)^2 = 804$ sq ft.)

In the present problem it is fairly obvious that $x = y = 25$ ft gives a maximum area since the area is always positive and its minimum is zero (for $x = 50$, $y = 0$ and for $x = 0$, $y = 50$). In other problems you may want to evaluate the function *in the neighborhood* of the obtained value of x to check whether it *gives a maximum or a minimum*. For example, for $x = 24$ ft and $y = 50 - 24 = 26$ ft, $A = 624$ sq ft, which is less than $A = 625$ sq ft. Hence, $x = y = 25$ ft gives a maximum.

It is worth noticing that a curve may have zero slope at a point and be neither maximum nor minimum there. For example, $y = x^3$ has:

$$\left.\frac{dy}{dx}\right]_{x=0} = 3x^2\,]_{x=0} = 0,$$

but is neither maximum nor minimum at $x = 0$. Figure 5.8 shows the behavior of $y = x^3$ around the origin O. The point O is called an *inflection point* of the curve. At an inflection point the tangent line (in this case, the x-axis) crosses the curve.

You can now solve directly the problem of Section 2.5 concerning the *minimum cost per floor* of a building. The total cost of the building for n floors was given by (2.5.2). Its cost per floor $c = C/n$ is then:

$$c = \frac{C}{n} = 0.04n^2 + 2n + 160 + \frac{2{,}000}{n}. \tag{c}$$

c is a function of n and its derivative with respect to n, dc/dn, must be zero when c is minimum. (The name of the function is now c and the name of the variable is n, while in the previous problem they were A and x, respectively. Notice that "name calling" does not make any difference.) Hence, differenti-

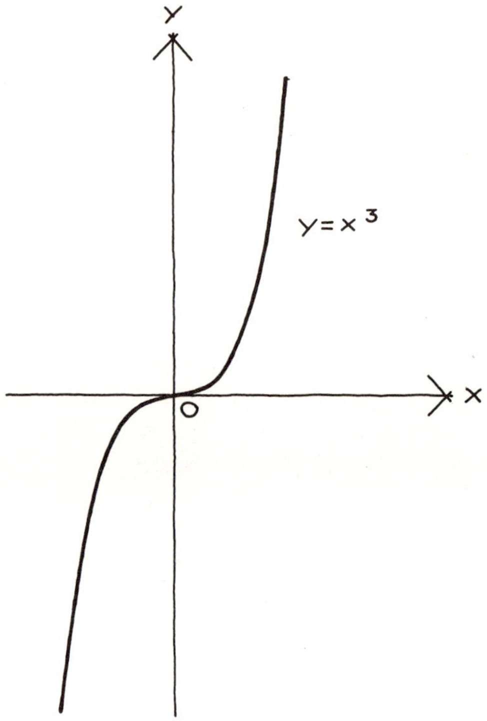

Figure 5.8

ating $c = f(n)$ we obtain the equation for n as:

$$\frac{dc}{dn} = 0.04 \times 2n + 2 - \frac{2{,}000}{n^2} = 0.08n + 2 - \frac{2{,}000}{n^2} = 0 \qquad \text{(d)}$$

or, multiplying through by n^2:

$$0.08n^3 + 2n^2 - 2{,}000 = 0. \qquad \text{(e)}$$

n may be obtained from (e) by the chords method of successive approximations of Section 4.5, i.e., by using (4.5.3) with, say, $n_1 = 10$, $n_2 = 30$ (this is done in problem 5.5.1). But the root n may also be evaluated directly from (d) by the following procedure of successive approximations, which you may want to learn because of its utter simplicity. Solve (d) for n^2:

$$0.08n + 2 = \frac{2{,}000}{n^2}; \quad \frac{n^2}{2{,}000} = \frac{1}{0.08n + 2}; \quad n^2 = \frac{2{,}000}{0.08n + 2},$$

and take the square root of both sides:

$$n = \sqrt{\frac{2{,}000}{0.08n + 2}}. \qquad \text{(f)}$$

(Since n must be positive we choose the $+$ sign in front of the square root.) Now give any approximate value n_0 to n on the right side of (f) and obtain a new value n_1 on the left. Use n_1 on the right and obtain a new value n_2 on the left, and keep repeating this process. The successive values $n_1, n_2, \ldots$ of n thus obtained are, in general, better and better approximations to the true value of n. [*Beware:* This may not be always true! But you can always check whether it is true or not by graphing the left-hand member of (e).] For example, starting with a value $n_0 = 10$ obtained graphically or through a judicious guess (you may always guess $n_0 = 0$, if you have no way of guessing a better value), we obtain:

$$n_1 = \sqrt{\frac{2{,}000}{0.08 \times 10 + 2}} = \sqrt{\frac{2{,}000}{2.80}} = \sqrt{714} = 27,$$

$$n_2 = \sqrt{\frac{2{,}000}{0.08 \times 27 + 2}} = \sqrt{\frac{2{,}000}{4.16}} = \sqrt{481} = 22,$$

$$n_3 = \sqrt{\frac{2{,}000}{0.08 \times 22 + 2}} = \sqrt{\frac{2{,}000}{3.76}} = \sqrt{532} = 23,$$

$$n_4 = \sqrt{\frac{2{,}000}{0.08 \times 23 + 2}} = \sqrt{\frac{2{,}000}{3.84}} = \sqrt{521} = 22.8 \cong 23.$$

Since n_3 equals n_4 within less than one unit and the number of floors n is an integer, $n = 23$ floors is the solution of the problem. For $n = 23$, (c) gives the value:

$$c = \frac{C}{n} = 0.04(23)^2 + 2 \times 23 + 160 + \frac{2{,}000}{23}$$
$$= 21 + 46 + 160 + 87 = 314 \quad \text{(or \$314,000)},$$

which agrees with the answer found in Section 2.5.

5.5.1 Solve (e) by chords method of successive approximations, using $n_1 = 10$, $n_2 = 30$ and calling y_1, y_2 the corresponding values of the left side of (e).

	0.08	2.0	0	−2,000
$n_1 = 10$		0.8	28	280
	0.08	2.8	28	$-1{,}720 = y_1$
	0.08	2.0	0	−2,000
$n_2 = 30$		2.4	132	3,960
	0.08	4.4	132	$+1{,}960 = y_2$

By the interpolation formula (4.5.3):

$$n_3 = \frac{1{,}960 \times 10 - (-1{,}720) \times 30}{1{,}960 - (-1{,}720)} = \frac{71{,}200}{3{,}680} = 19.$$

Check that by substituting in (4.5.3) $n_3 = 19$, $y_3 = -727$ for $n_1 = 10$, $y_1 = -1{,}720$, one obtains $n_4 = 22$, $y_4 = -180$ and similarly $n_5 = 22.7 \simeq 23$.

It is seen that in this case the chords method is slower than the preceding method in solving for n.

Find the maximum and minimum values of the following functions by setting their derivatives equal to zero, and check whether the values are maxima or minima.

5.5.2 $y = x^4 - 1.$ $(x = 0, \text{min.})$

5.5.3 $y = x^4 - x^2 + 2.$ $\left(x = +\frac{1}{\sqrt{2}}, \text{min.}; \quad x = -\frac{1}{\sqrt{2}}, \text{min.}; x = 0, \text{max.}\right)$

5.5.4 $y = x^4 - x^3 + x.$ $(x = -0.46, \text{min.})$

5.5.5 $y = \sin x;$ $-\pi \leqq 0 \leqq \pi.$ $\left(x = \frac{\pi}{2}, \text{max.}\right)$

5.5.6 $y = x^4 + \frac{2}{x^2} + 3.$ $\left(x = 1, \text{min.}\right)$

5.6 THE RATE OF CHANGE OF A FUNCTION

You were introduced to the derivative as the slope of a curve so that you could visualize the limiting process used in the evaluation of the derivative, which is symbolized by the last step of the delta process:

$$\lim_{\Delta x \to 0} \frac{\Delta y_1}{\Delta x_1}.$$

But you may have already noticed that the "slope" of some of the curves used in our problems had a wider significance than that of a geometrical slope. For example, in the building cost problem the "slope" is measured in "dollars-per-floor per floor."

This wider significance of the derivative can also be visualized and used in approximate calculations. In Fig. 5.9 as one moves from P_1 to P along the curve y, i.e., as x moves along the x-axis from $x = x_1$ to $x = x_1 + \Delta x$, y grows from $y = y_1$ to $y = y_1 + \Delta y$, i.e., by $AP = \Delta y$.

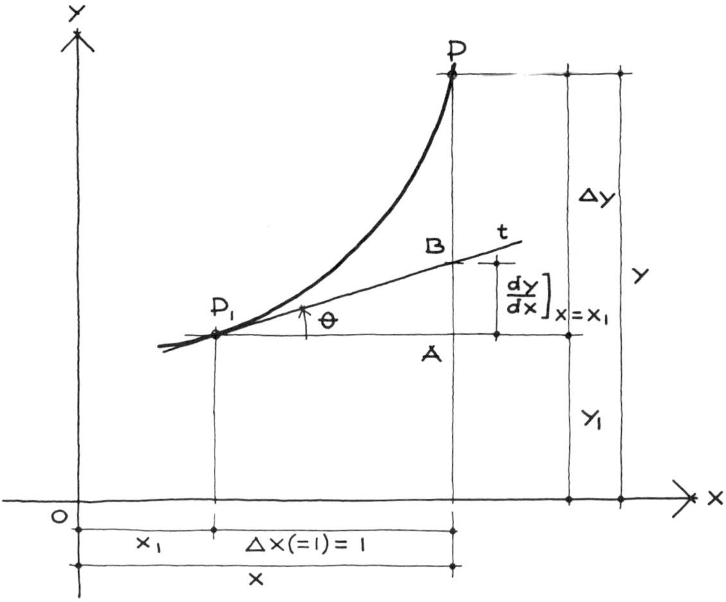

Figure 5.9

Instead of moving along the curve y, let us now move along the tangent line t at P_1. As x goes from $x = x_1$ to $x = x_1 + \Delta x$, y now grows only by AB.

In the triangle P_1BA:

$$\frac{AB}{P_1A} = \frac{AB}{\Delta x} = \tan\theta = m_1 = \frac{dy}{dx}\bigg]_{x=x_1},$$

where m_1 is the slope of y at P_1. Hence, multiplying both sides by $P_1A = \Delta x$:

$$AB = \frac{dy}{dx}\bigg]_{x=x_1} \Delta x. \tag{a}$$

For a change in Δx equal to 1 (say, one inch):

$$AB = \frac{dy}{dx}\bigg]_{x=x_1} \times 1 = \frac{dy}{dx}\bigg]_{x=x_1} \tag{b}$$

Equation (b) shows that the derivative $dy/dx]_{x=x_1}$ measures also the change in *y along the tangent line* due to a unit change in x. In other words, if, from P_1 on, the curve y had a *constant* slope equal to m_1, the change in y due to a unit change in x would be equal to $dy/dx]_{x=x_1}$. The change in a function due to a unit change in the variable is called the *rate of change* of the function. Hence, *the derivative at a given point measures the rate of change of the function at that point.*

For example, let us assume that, as you drive away from New York, you plot your distance s from the city versus the time t, measured from the moment you start traveling (Fig. 5.10). The slope $ds/dt\,]_{t=t_1}$ in Fig. 5.10 then measures the rate of change of your distance with time, which is what we call *speed.* When you say: "Right now my car is doing 55 miles an hour," you mean to say: "If from now on I kept the speed of my car exactly constant, and if I drove 1 hour, my car would cover 55 miles." By the shorthand of the calculus your statement becomes: "At this time, say $t = t_1$, the derivative of the function $s = f(t)$ equals 55."

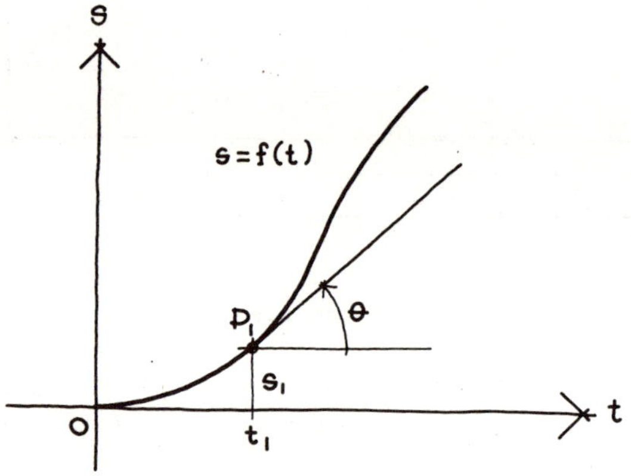

Figure 5.10

You may now make efficient use of the derivative to evaluate *small* changes in y due to *small* changes in x. From Fig. 5.9 you see that if the change Δx in x becomes smaller and smaller the difference between AP and AB also becomes smaller and smaller. If Δx is small enough, "going off on the tangent" at P_1 rather than staying on the curve does not introduce a substantial error in Δy.

The change AB along the tangent is so useful that it has a name and is indicated by the special symbol dy, whenever Δx is small. By (a):

$$AB \equiv dy = \frac{dy}{dx}\bigg]_{x=x_1} \Delta x. \tag{5.6.1}$$

dy is called *the differential of y at $x = x_1$*. The symbol dy must not be confused with the algebraic product of d times y, but is a calculus symbol that gives the order: "Multiply the derivative of y at $x = x_1$ by a (small) Δx."

Let us measure the difference between the change Δy along the curve and the change dy along the tangent in a simple case. The area A of a square plot of side x equals x^2. By how much does A change if the side changes from $x = 100$ ft to $x + \Delta x = 102$ ft?

Calling ΔA the change in A:

$$A = f(x) = x^2,$$

$$A + \Delta A = f(x + \Delta x) = (x + \Delta x)^2 = x^2 + 2x\,\Delta x + (\Delta x)^2,$$

and by subtraction:

$$\Delta A = 2x\,\Delta x + (\Delta x)^2. \qquad \text{(c)}$$

On the other hand, the derivative of A is:

$$\frac{dA}{dx} = \frac{dx^2}{dx} = 2x,$$

so that by (5.6.1):

$$dA = \frac{dA}{dx}\,\Delta x = 2x\,\Delta x. \qquad \text{(d)}$$

The difference between ΔA and dA is the difference between (c) and (d):

$$\Delta A - dA = 2x\,\Delta x + (\Delta x)^2 - 2x\,\Delta x = (\Delta x)^2$$

and is seen to become small relative to $2x\,\Delta x$ as soon as Δx is small in comparison with $2x$.

In our problem:

$$A = (100)^2 = 10{,}000 \text{ sq ft.}$$
$$A + \Delta A = (102)^2 = 10{,}404 \text{ sq ft.}$$
$$\Delta A = 404 \text{ sq ft,}$$
$$dA = 2 \times 100 \times 2 = 400 \text{ sq ft,}$$
$$\Delta A - dA = 404 - 400 = 4 \text{ sq ft,}$$

which is only 1% of ΔA. Figure 5.11 illustrates this result graphically.

You may now estimate by the differential the change in y at $x = x_1$ due to a given Δx (small in comparison with x_1) for the following functions.

5.6.1 $y = \sqrt{x}$, $x_1 = 1$, $\Delta x = 0.1$. ($dy = 0.05$.)

5.6.2 $y = \sin x$, $x_1 = \pi/2$, $\Delta x = 0.1$. ($dy = 0$.)

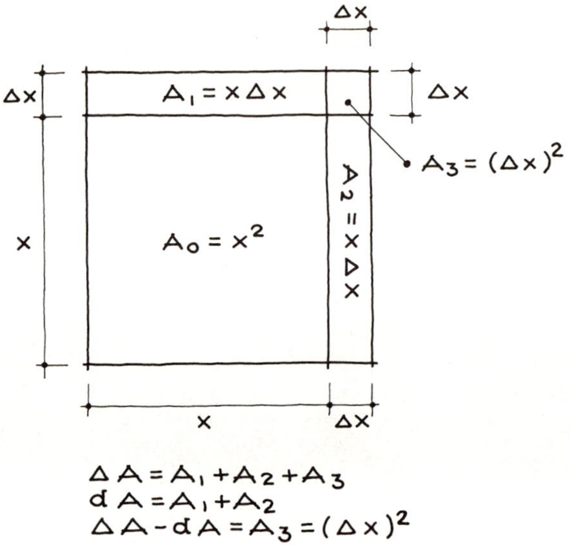

Figure 5.11

5.6.3 $y = 1/x$, $x_1 = 2$, $\Delta x = 0.2$. $\quad (dy = -0.05.)$

5.6.4 $y = \tan x$, $x_1 = 1.5$, $\Delta x = 0.1$. $\quad (dy = 20.6.)$

5.6.5 $y = x^2 - 2/x$, $x_1 = 20$, $\Delta x = 1$. $\quad (dy = 40.005.)$

5.6.6 $y = 1/x^3 + 1/x^2$, $x_1 = 1$, $\Delta x = 0.2$. $\quad (dy = -1.)$

You may have been puzzled by the symbol dy/dx for the derivative. The differential helps justify this symbol, used first by Leibnitz, a German mathematician and philosopher, who invented the calculus at the same time as Newton, but independently. Just as dy indicates a small change in y, dx is used to indicate a smaller and smaller Δx. If we use dx for a small Δx in the definition of dy [equation (5.6.1)]:

$$dy = (\text{derivative of } y \text{ at } x = x_1)\, dx,$$

and divide both sides by dx, we obtain:

$$\left.\frac{dy}{dx}\right]_{x=x_1} = \text{derivative of } y \text{ at } x = x_1.$$

The Leibnitz symbol for derivative indicates that dy/dx may be also considered as a fraction, in which both numerator and denominator become smaller together.

5.7 DIFFERENTIATION SHORTCUTS

Although the delta method allows the evaluation of the derivative of *any* function, in practice it is essential to learn the following three shortcuts in order to save time and avoid mistakes.

(a) *Derivative of the product of two functions*

If u and v are two functions of x:

$$u = f_1(x), \quad v = f_2(x), \tag{5.7.1}$$

and $y = F(x)$ is a function equal to the product of u times v:

$$y = F(x) = f_1(x) \times f_2(x) = u \times v, \tag{5.7.2}$$

applying the delta method to u, v and y, we obtain:

$$\begin{aligned}
f_1(x_1 + \Delta x) &= u + \Delta u, \quad f_2(x_1 + \Delta x) = v + \Delta v, \\
F(x_1 + \Delta x) &= f_1(x_1 + \Delta x) \times f_2(x_1 + \Delta x), \\
\Delta y &= F(x_1 + \Delta x) - F(x_1) \\
&= f_1(x_1 + \Delta x) \times f_2(x_1 + \Delta x) - f_1(x_1) f_2(x_1) \\
&= (u + \Delta u)(v + \Delta v) - uv \\
&= \cancel{uv} + u\,\Delta v + v\,\Delta u - \cancel{uv} = u\,\Delta v + v\,\Delta u, \\
\frac{\Delta y}{\Delta x} &= u\frac{\Delta v}{\Delta x} + v\frac{\Delta u}{\Delta x},
\end{aligned}$$

$$\frac{dy}{dx} = \lim_{\Delta x \to 0} \frac{\Delta y}{\Delta x} = u \lim_{\Delta x \to 0} \frac{\Delta v}{\Delta x} + v \lim_{\Delta x \to 0} \frac{\Delta u}{\Delta x} = u\frac{dv}{dx} + v\frac{du}{dx}. \tag{5.7.3}$$

In words: The derivative of the product of two functions equals the first function times the derivative of the second plus the second times the derivative of the first. For example, given:

$$\begin{aligned}
y &= x \sin x, \\
u &= x, \quad \frac{du}{dx} = 1, \\
v &= \sin x, \quad \frac{dv}{dx} = \cos x, \\
\frac{dy}{dx} &= u\frac{dv}{dx} + v\frac{du}{dx} = x \cos x + \sin x.
\end{aligned}$$

Check the following derivatives by the multiplication rule.

5.7.1 $y = x^2 \cos x$ $\left(\frac{dy}{dx} = -x^2 \sin x + 2x \cos x\right)$.

5.7.2 $y = x^2 \tan x$ $\left(\frac{dy}{dx} = 2x \tan x + \frac{x^2}{\cos^2 x}\right)$.

5.7.3 $y = \frac{\sin x}{x}$ $\left(\frac{dy}{dx} = -\frac{\sin x}{x^2} + \frac{\cos x}{x}\right)$.

$\left(\textit{Hint: } \frac{1}{x} = x^{-1}.\right)$

5.7.4 $y = 2 \sin x \cos x + 1 \left(\frac{dy}{dx} = 2(\cos^2 x - \sin^2 x)\right)$.

5.7.5 $y = (\sin x + \cos x)^2$ $\left(\frac{dy}{dx} = 2(\cos^2 x - \sin^2 x)\right)$.

Hint: Expand the square $(\sin x + \cos x)^2$ first.

Note: The functions of Problems 5.7.4 and 5.7.5 have the same derivative, although they are not identical. Find out why by plotting their graphs.

(b) *Derivative of the ratio of two functions*

If:

$$u = f_1(x), \quad v = f_2(x),$$

and:

$$y = F(x) = \frac{f_1(x)}{f_2(x)} = \frac{u}{v}, \tag{5.7.4}$$

we obtain by the delta method:

$$f_1(x_1 + \Delta x) = u + \Delta u, \quad f_2(x_1 + \Delta x) = v + \Delta v,$$

$$F(x_1 + \Delta x) = \frac{f_1(x_1 + \Delta x)}{f_2(x_1 + \Delta x)} = \frac{u + \Delta u}{v + \Delta v},$$

$$\Delta y = F(x_1 + \Delta x) - F(x_1) = \frac{u + \Delta u}{v + \Delta v} - \frac{u}{v},$$

or, reducing the two fractions to the same denominator:

$$\Delta y = \frac{v(u + \Delta u) - u(v + \Delta v)}{v(v + \Delta v)} = \frac{\cancel{vu} + v\,\Delta u - \cancel{uv} - u\,\Delta v}{v(v + \Delta v)}$$

$$= \frac{v\,\Delta u - u\,\Delta v}{v(v + \Delta v)}.$$

Dividing both sides by Δx, i.e., dividing on the left Δy and on the right each term of the *numerator* of the fraction by Δx, $\Delta y/\Delta x$ becomes:

$$\frac{\Delta y}{\Delta x} = \frac{v\,\Delta u/\Delta x - u\,\Delta v/\Delta x}{v(v + \Delta v)}.$$

As $\Delta x \longrightarrow 0$, v and u remain constant but Δv approaches zero so that the denominator becomes equal to v^2 and:

$$\frac{dy}{dx} = \lim_{\Delta x \to 0} \frac{v\,\Delta u/\Delta x - u\,\Delta v/\Delta x}{v(v + \Delta v)} = \frac{v \lim_{\Delta x \to 0} \Delta u/\Delta x - u \lim_{\Delta x \to 0} \Delta v/\Delta x}{v^2},$$

or:

$$\frac{dy}{dx} = \frac{v\,du/dx - u\,dv/dx}{v^2}. \tag{5.7.5}$$

For example, with:

$$y = \frac{\sin x}{\cos x} = \tan x,$$

$$\frac{dy}{dx} = \frac{\cos x\,(\cos x) - \sin x\,(-\sin x)}{\cos^2 x} = \frac{\cos^2 x + \sin^2 x}{\cos^2 x} = \frac{1}{\cos^2 x},$$

as was found by the delta method in Section 5.4 [Eq. (5.4.3)].

You may now check the following derivatives.

5.7.6 $y = \dfrac{\cos x}{\sin x} \quad \left(\dfrac{dy}{dx} = -\dfrac{1}{\sin^2 x}\right).$

5.7.7 $y = \dfrac{\tan x}{x^2} \quad \left(\dfrac{dy}{dx} = \dfrac{x - 2\sin x \cos x}{x^3 \cos^2 x}\right).$

5.7.8 $y = \dfrac{x^2}{\sin x} \quad \left(\dfrac{dy}{dx} = \dfrac{x(2\sin x - x\cos x)}{\sin^2 x}\right).$

5.7.9 $y = \dfrac{x^2 + 1}{x^3 + 4} \quad \left(\dfrac{dy}{dx} = \dfrac{-x^4 - 3x^2 + 8x}{(x^3 + 4)^2}\right).$

(c) *The chain rule*

So far we have differentiated simple functions of x. One of the most important shortcuts in the differentiation of complicated functions of x is obtained by introducing an *intermediate variable*. For example, let us assume that the cost C of a building consists of a constant factor a and a factor bn^2 proportional to the square of the number of floors n:

$$C = a + bn^2. \tag{a}$$

Let us also assume that the number of floors needed to house N people consists of a constant factor c and a factor eN proportional to N:

$$n = c + eN. \tag{b}$$

By (a) and (b) the cost to house N people is:

$$C = a + b(c + eN)^2 = a + b(c^2 + 2ceN + e^2N^2).$$

The rate of change of C with N, i.e., the increase in cost to house one more person above N persons (see Section 5.6) is:

$$\frac{dC}{dN} = 2bce + 2be^2N = 2be(c + eN). \tag{c}$$

The same rate of change may be obtained directly by evaluating separately the two derivatives:

$$\frac{dC}{dn} = 2bn, \qquad \frac{dn}{dN} = e,$$

multiplying one by the other, and making use of (b):

$$\frac{dC}{dN} = \frac{dC}{dn} \times \frac{dn}{dN} = 2ben = 2be(c + eN). \tag{d}$$

This rule, the so-called *chain rule* of differentiation, may be proved to hold in general. Let y be a function of a variable z (n in the previous example) and z be a function of a variable x (N in the previous example):

$$y = f_1(z), \qquad z = f_2(x), \tag{5.7.6}$$

so that y is a function $F(x)$ of x "through z":

$$y = f_1(z) = f_1[f_2(x)] \equiv F(x).$$

Let us write the ratio $\Delta y/\Delta x$, and multiply and divide this ratio by Δz:

$$\frac{\Delta y}{\Delta x} = \frac{\Delta y}{\Delta x}\frac{\Delta z}{\Delta z} = \frac{\Delta y}{\Delta z}\frac{\Delta z}{\Delta x}.$$

If we let $\Delta x \longrightarrow 0$, as we do in the last step of the delta method, Δz will also approach zero since z is a function of x. But if Δz approaches zero, Δy will also approach zero since y is a function of z. Hence:

$$\frac{dy}{dx} = \lim_{\Delta x \to 0} \frac{\Delta y}{\Delta x} = \lim_{\Delta z \to 0} \frac{\Delta y}{\Delta z} \times \lim_{\Delta x \to 0} \frac{\Delta z}{\Delta x}$$

or:

$$\frac{dy}{dx} = \frac{dy}{dz} \times \frac{dz}{dx}. \tag{5.7.7}$$

The chain rule of (5.7.7) is the most important shortcut in differentiation: practically, it would be very hard to differentiate without it. For example, let:

$$y = (1 + 3x^2)^2$$

and set:

$$(1 + 3x^2) = z,$$

so that:

$$\frac{dz}{dx} = 6x.$$

Then, by (5.7.7) with:

$$z = 1 + 3x^2, \qquad y = z^2, \quad \frac{dy}{dz} = 2z,$$

$$\frac{dy}{dx} = 2z \times 6x = 12xz = 12x(1 + 3x^2).$$

This result may be checked by expanding y:

$$y = (1 + 3x^2)^2 = 1 + (2 \times 1 \times 3x^2) + (3x^2)^2 = 1 + 6x^2 + 9x^4$$

$$\frac{dy}{dx} = 12x + 36x^3 = 12x(1 + 3x^2).$$

The following derivative, instead, could only be checked by the five steps of the delta method:

$$y = \sin(x^2), \quad z = x^2, \quad y = \sin z,$$

$$\frac{dy}{dx} = \frac{d \sin z}{dz}\frac{dz}{dx} = \cos z \times 2x = 2x \cos(x^2).$$

Try your hand at the following derivatives.

5.7.10 $y = \sin(2x - 1)$ $\left(\frac{dy}{dx} = 2\cos(2x - 1)\right).$

5.7.11 $y = \cos^2 x$ $\left(\frac{dy}{dx} = -2\sin x \cos x\right).$

(*Hint:* $\cos^2 x$ means $(\cos x)^2$; let $\cos x = z$).

5.7.12 $y = \sqrt{\cos x + 2}$ $\left(\frac{dy}{dx} = \frac{-\frac{1}{2}\sin x}{\sqrt{\cos x + 2}}\right).$

5.7.13 $y = (1 + \sin x)^{3/2}$ $\left(\frac{dy}{dx} = \frac{3}{2}(1 + \sin x)^{1/2}\cos x\right).$

5.7.14 $y = \frac{1}{\cos^2 x}$ $\left(\frac{dy}{dx} = \frac{2\sin x}{\cos^3 x}\right).$

Hint: $\frac{1}{\cos^2 x} = (\cos x)^{-2}.$

The chain rule may be applied successively more than once. For example, if:

$$y = \sqrt{\cos 2x},$$

let:

$$2x = w, \qquad \cos 2x = \cos w = z.$$

Then:

$$\frac{dy}{dx} = \frac{dy}{dz} \times \frac{dz}{dw} \times \frac{dw}{dx}, \tag{5.7.8}$$

so that:

$$y = \sqrt{z} = z^{1/2}, \qquad \frac{dy}{dz} = \frac{1}{2}z^{-1/2};$$

$$z = \cos w, \qquad \frac{dz}{dw} = -\sin w;$$

$$w = 2x, \qquad \frac{dw}{dx} = 2;$$

$$\frac{dy}{dx} = \frac{1}{2}z^{-1/2}(-\sin w)(2) = -\frac{\sin 2x}{\sqrt{\cos 2x}}.$$

Use the chain rule on the following functions.

5.7.15 $y = \tan\left(\frac{1}{x^2 + 1}\right)$ $\left(\frac{dy}{dx} = -\frac{2x}{(x^2 + 1)^2}\frac{1}{\cos^2[1/(x^2 + 1)]}\right).$

5.7.16 $y = (1 + \sin 3x^2)^{3/2}$ $\left(\frac{dy}{dx} = 9x\cos 3x^2(1 + \sin 3x^2)^{1/2}\right).$

6

INTEGRAL CALCULUS

6.1 INTRODUCTION

The basic concept of derivative was introduced visually in Chapter 5 by noticing that the *derivative* is geometrically equivalent to the *slope of a curve*.

The second basic concept of the calculus, the *integral*, is geometrically equivalent to the *area bounded by a curve*. Moreover, it is strictly related to the concept of derivative and plays a very important role in evaluating not only areas, but also, for example, volumes, geometric properties of structural sections, and amounts of heat and flow of air in environmental systems, just to mention a few fields where you will find integrals of great practical value.

6.2 THE AREA BOUND BY A POLYGON

The area A_r of a rectangle of sides b, h (Fig. 6.1) is:

$$A_r = b \times h. \qquad (6.2.1)$$

The area A_t of a right triangle of sides b, h (Fig. 6.1) is half the area

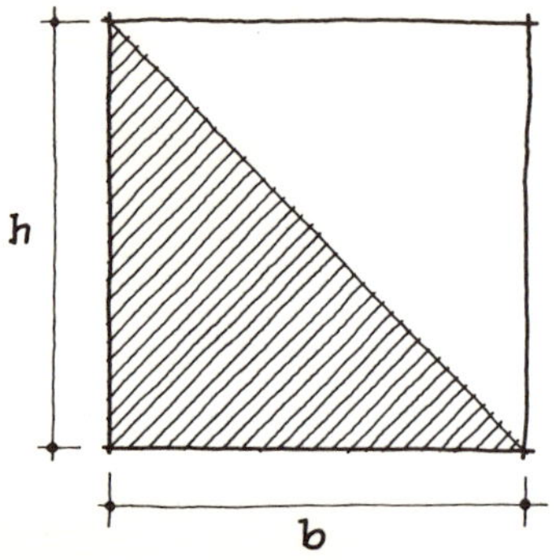

Figure 6.1

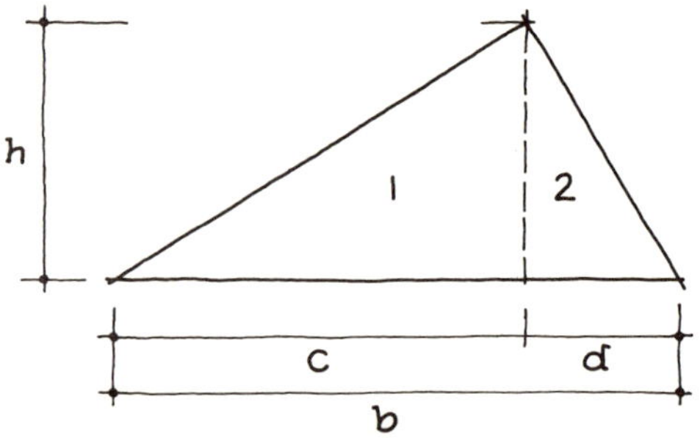

Figure 6.2

of the rectangle of sides b, h:

$$A_t = \tfrac{1}{2}b \times h. \tag{6.2.2}$$

The area of any triangle may be obtained as the sum of the areas of two right triangles 1, 2 (Fig. 6.2) with the same height h:

$$A_t = \tfrac{1}{2}c \times h + \tfrac{1}{2}d \times h = \tfrac{1}{2}(c + d)h = \tfrac{1}{2}bh. \tag{6.2.2a}$$

The area of any polygon may be obtained by summing the areas of triangles with a common vertex V inside the polygon (Fig. 6.3).

The area of a *regular polygon* with n sides, inscribed in a circle of radius R and center O (Fig. 6.4a), is the sum of the areas of n identical triangles with an angle α at O equal to $2\pi/n$ radians. The height and base

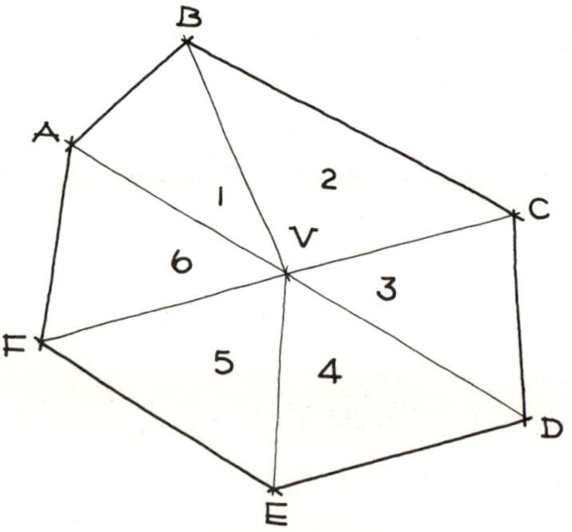

Figure 6.3

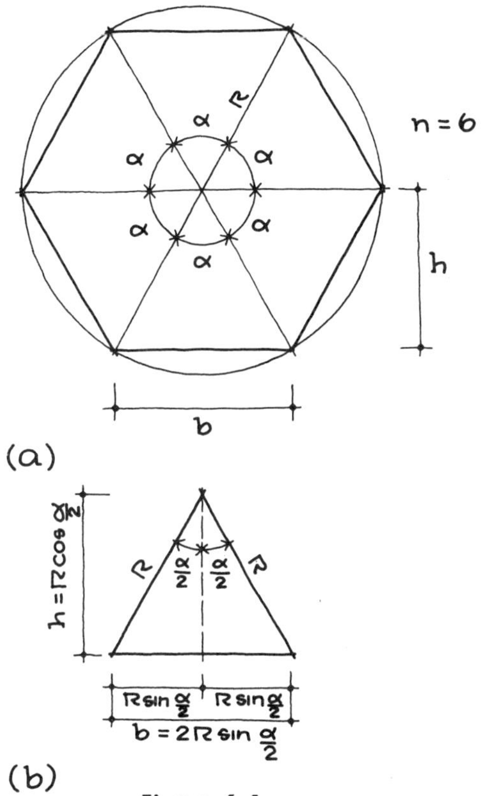

Figure 6.4

of these triangles are obtained by trigonometry (Fig. 6.4b):

$$h = R\cos\frac{\alpha}{2} \qquad b = 2R\sin\frac{\alpha}{2}\cdot$$

Hence, the area of one triangle is:

$$A_1 = \frac{1}{2}\left(2R\sin\frac{\alpha}{2}\right)\left(R\cos\frac{\alpha}{2}\right) = \frac{1}{2}R^2\left(2\sin\frac{\alpha}{2}\cos\frac{\alpha}{2}\right),$$

or, by (3.4.11):

$$A_1 = \frac{1}{2}R^2\sin 2\left(\frac{\alpha}{2}\right) = \frac{1}{2}R^2\sin\alpha = \frac{1}{2}R^2\sin\frac{2\pi}{n}\cdot$$

The area A_n of the polygon is the area of n triangles:

$$A_n = nA_1 = \left(\frac{n}{2}\sin\frac{2\pi}{n}\right)R^2. \tag{6.2.3}$$

For example, for $n = 6$ we obtain:

$$\frac{2\pi}{6} = \frac{360^\circ}{6} = 60^\circ, \qquad \sin 60^\circ = \frac{\sqrt{3}}{2}$$

$$A_6 = \frac{6}{2}\frac{\sqrt{3}}{2}R^2 = \frac{3\sqrt{3}}{2}R^2 = 2.60R^2$$

Table 6.1 gives the area of regular polygons with an increasing number of sides n, divided by R^2.

Table 6.1

n	A_n/R^2	n	A_n/R^2	n	A_n/R^2
3	1.30	15	3.05	60	3.136
6	2.60	18	3.08	120	3.140
9	2.89	24	3.106	360	3.141
12	3.00	30	3.119		

6.3 THE AREA OF THE CIRCLE

If we try to compute the area of the circle of radius R, we run for the first time into the problem of evaluating the area bound by a curve rather than by the straight sides of a polygon. To evaluate the area of the circle we shall follow the same procedure we used in determining the slope of a curve: successive approximations leading to a limiting process.

It is visually obvious that, as the number of sides of the regular polygon inscribed in the circle increases, the area of the polygon becomes nearer and nearer to the area of the circle. If we scan Table 6.1, we see that as n increases there is a trend in the coefficient of R^2: it approaches (to two decimal figures) 3.14. Hence, we accept the area of the circle A_c to be *approximately*:

$$A_c \simeq 3.14R^2. \tag{6.3.1}$$

The mathematical shorthand symbol to state that n becomes bigger and bigger is:

$$\lim_{n\to\infty}$$

(read "limit as n approaches infinity"), by means of which the area A_c of the circle may be written *exactly*, taking the limit of (6.2.3), as:

$$A_c = \lim_{n\to\infty} \left(\frac{n}{2} \sin \frac{2\pi}{n}\right) R^2. \tag{6.3.2}$$

The value of $\sin x$ for very small values of x given by (3.7.1) allows us to find the exact value of the limit in (6.3.2). By (3.7.1), as x approaches zero, $\sin x \simeq x$. Hence, as n grows, and consequently $2\pi/n$ approaches zero, $\sin 2\pi/n \simeq 2\pi/n$ and (6.3.2) becomes:

$$A_c = \left(\frac{n}{2}\right)\left(\frac{2\pi}{n}\right) R^2 = \pi R^2. \tag{6.3.3}$$

The correct value of the coefficient of R^2 in the area of the circle is $\pi = 3.14159\ldots$ (a so-called irrational number, whose infinite decimal figures never repeat, but follow each other at random).

It is important to realize that we have obtained the area bound by a circle *as the sum of the areas of an increasing number of triangles of decreasing area.* We cannot, of course, take an "infinite number of triangles of zero area," but we can follow the trend of the sum as n grows and, consequently, as the area of each triangle decreases. The operation of taking the sum of an increasing number of smaller and smaller areas to obtain the area bound by a curve is typical of a mathematical process called *integration.* The process of integration is somewhat analogous to the process of differentiation, because both are limiting processes: in differentiation Δx, and hence Δy, approach zero; in integration the areas of the single triangles, or, more generally, of the single elements of area approach zero.

In the next section we shall extend the concept of integration in order to increase its usefulness.

6.4 THE PROCESS OF INTEGRATION AS A SUM

The cost c of each vertical foot of a building depends on its elevation h above ground and consists of a constant factor and of a factor proportional to the square of the elevation:

$$c = a + bh^2. \tag{6.4.1}$$

How much will a building H ft high cost?

Since each vertical foot of building has a different cost, one may obtain a *rough* idea of the total cost of the building by assuming, instead, that the cost per foot remains *constant* over, say, each 50 ft of height. For example, with:

$$a = 60{,}000 \text{ \$/ft}, \quad b = 1.6 \text{ \$/cu ft}, \quad H = 200 \text{ ft},$$

(6.4.1) becomes:

$$c(h) = 60{,}000 + 1.6h^2$$

and at $h = 50, 100, 150, 200$ ft the cost per foot is:

h (ft)	c (\$/ft)
50	64,000
100	76,000
150	96,000
200	124,000

Adding up the cost of each 50-ft section of constant cost per foot, the approximate total cost C is roughly:

$$\begin{aligned} C &= 50 \times 64{,}000 + 50 \times 76{,}000 + 50 \times 96{,}000 + 50 \times 124{,}000 \\ &= 50(64{,}000 + 76{,}000 + 96{,}000 + 124{,}000) \\ &= 50 \times 360{,}000 = \$18{,}000{,}000. \end{aligned}$$

A better approximation of C may be obtained by subdividing the building into 8 sections, each 25 ft high, and assuming c to be constant for each 25-ft section. With the following additional values of c at $h = 25, 75, 125$ and 175 ft:

h (ft)	c (\$/ft)
25	61,000
75	69,000
125	85,000
175	109,000

we obtain:

$$\begin{aligned} C = 25(&61{,}000 + 64{,}000 + 69{,}000 + 76{,}000 + 85{,}000 + 96{,}000 \\ &+ 109{,}000 + 124{,}000) \\ = 25 &\times 684{,}000 = \$17{,}100{,}000. \end{aligned}$$

It takes a little longer to evaluate the cost per foot every 10 ft and add 20 terms to find the better approximation $C = \$16{,}592{,}000$. It takes a desk

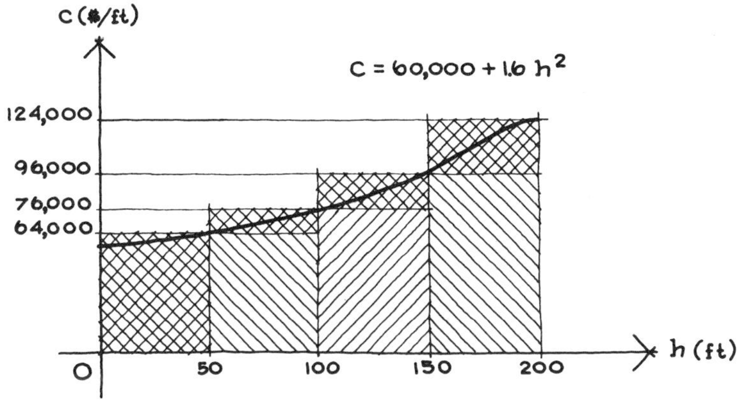

Figure 6.5

calculator to evaluate the cost at every foot and add 200 terms to find $C = \$16{,}298{,}720$.

This process of successive approximations may be visualized in Fig. 6.5, in which c is plotted versus h in a Cartesian graph. The approximation $C = \$18{,}000{,}000$ is obtained by adding the areas of the 4 rectangles of width 50 ft in Fig. 6.5; the next approximation by adding the areas of 8 rectangles of width 25 ft, etc. Thus, what we are trying to obtain is the exact value of the area in Fig. 6.5 which is bound by the graph of c versus h, the y-axis, the x-axis and the vertical through $h = 200$ ft, as the sum of an increasing number of rectangles of decreasing width.

Since we have taken the height of the rectangles at the *right* end of their base, in this problem the approximations we obtain are always too high, i.e., from above. By taking the rectangle heights at the *left* end of their base (Fig. 6.5), we obtain, instead, the following approximations, which are too low, i.e., from below:

With 4 sections:

$$\begin{aligned} C &= 50(0 + 64{,}000 + 76{,}000 + 96{,}000) \\ &= 50 \times 236{,}000 = \$11{,}800{,}000. \end{aligned}$$

With 8 sections:

$$\begin{aligned} C &= 25(0 + 61{,}000 + 64{,}000 + 69{,}000 + 76{,}000 + 85{,}000 \\ &\quad + 96{,}000 + 109{,}000) \\ &= 25 \times 560{,}000 = \$14{,}000{,}000. \end{aligned}$$

Similarly, for 20 terms $C = \$15{,}952{,}000$ and for 200 terms $C = \$16{,}234{,}720$.

Table 6.2

Δh (ft)	C_a(\$)	C_b(\$)
50	18,000,000	11,800,000
25	17,100,000	14,000,000
10	16,592,000	15,952,000
1	16,298,720	16,234,720

A perusal of Table 6.2 shows that as Δh becomes smaller there is a trend towards the same value in the approximations C_a from above and in those C_b from below, and that the area under the curve c between $h = 0$ and $h = 200$ ft must be given with good accuracy by the average of C_a and C_b, for $\Delta h = 1$, that is, \$16,266,720.

The procedure used to obtain the area under the curve $c(h)$ may be obviously generalized to establish the cost of a building of *any* height H. All that we have to do is add the areas of the rectangles up to H, where H may be 200 ft as in the present example, or may have any other value.

Let us now write the various steps of the integration process for a general curve $y = f(x)$, just as we did in differentiation for the delta method. In Fig. 6.6 the area A to be evaluated is bound by the curve $y = f(x)$, the x-axis and the verticals at $x = a$ and $x = b$. We call Δx the constant width of the rectangles and n their number ($n = 6$ in Fig. 6.6). We then perform the following operations:

(a) Evaluate the function y at the right end of the intervals of width Δx,

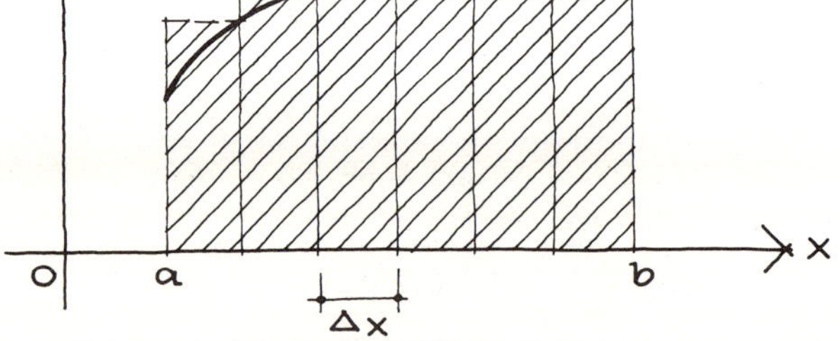

Figure 6.6

i.e., compute the n ordinates or heights:

$$f(a + \Delta x),\quad f(a + 2\,\Delta x),\quad f(a + 3\,\Delta x), \ldots, f(a + n\,\Delta x).$$

(b) Multiply each ordinate by $\Delta x = (b - a)/n$, to obtain the areas of the rectangles, and add up all the products:

$$f(a + \Delta x)\,\Delta x + f(a + 2\,\Delta x)\,\Delta x + \ldots + f(a + n\,\Delta x)\,\Delta x.$$

(c) Make Δx smaller and smaller, i.e., take the limit as Δx approaches zero and, consequently, n becomes larger and larger, i.e., approaches infinity.

These steps are conveniently expressed by some additional mathematical shorthand symbols. Instead of writing $f(a + \Delta x)$, $f(a + 2\,\Delta x)$, $f(a + 3\,\Delta x), \ldots$, let us indicate these ordinates by $f_1, f_2, f_3, \ldots$, where the subscripts $1, 2, 3, \ldots$, show how many times Δx must be added to a [in this shorthand, for example, $f_n = f(a + n\,\Delta x) = f(b)$]. Then, the 3 steps of the process are symbolized by:

$$A = \lim_{n\to\infty} (f_1 + f_2 + f_3 + \ldots + f_n)\,\Delta x.$$

The sum of n values of f, subscripted from 1 to n, is usually indicated by another shorthand symbol using the Greek letter $\sum$ (read "sigma"), which stands for our S and means "the sum of":

$$A = \lim_{n\to\infty} \sum_{i=1}^{n} f_i\,\Delta x. \tag{6.4.1}$$

The $\sum$ symbol for summation tells you to add the products of Δx times the values of the function $f(x)$ at $x = a + i\,\Delta x$, indicated by the subscript i, where i is first 1, then 2, then $3, \ldots$, and finally n, as shown by the fact that i starts at 1 under $\sum$ and ends at n above $\sum$. For example:

$$\sum_{i=3}^{6} f_i\Delta x = (f_3 + f_4 + f_5 + f_6)\,\Delta x = [f(a + 3\,\Delta x) + f(a + 4\,\Delta x) + f(a + 5\,\Delta x) + f(a + 6\,\Delta x)]\,\Delta x.$$

A last shorthand symbol is finally introduced to indicate the limiting process in the evaluation of A. Just as in the case of the derivative, a Δx approaching zero will be indicated by dx (see Section 5.6) and the limit of the sum as n grows to infinity will be indicated by a capital S (for sum), which traditionally is a very "lean" S or $\int$, with the value of a at the bottom and the value of b at the top to indicate the *limits* of the values of x in the sum. Moreover, $f(x)$ takes the place of the subscripted values f_i. Thus:

$$A \equiv \int_a^b f(x)\,dx \equiv \lim_{n\to\infty} \sum_{i=1}^{n} f_i\,\Delta x. \tag{6.4.2}$$

You may want to think of A as the sum of an infinite number of rectangles of practically zero width, whose heights are the values of $f(x)$ at *all* the values of x between the *lower limit* $x = a$ and the *upper limit* $x = b$.

Sometimes you may need to evaluate the area A between a *fixed* lower limit $x = a$ and an indefinite or *variable* upper limit $x = X$, where X may equal b in one case, c in another, d in another (just as the height H of the building in our example was 200 ft, but could be 300 ft or 150 ft in other cases). To indicate a variable upper limit of integration X you must substitute X for b:

$$A = \int_a^X f(x)\,dx. \tag{6.4.3}$$

The 3 steps in the evaluation of A are called the *integration of the function* $f(x)$. The symbol:

$$\int_a^b f(x)\,dx$$

(read "integral from a to b of $f(x)\,dx$") is called the *definite integral of* $f(x)$ *between* a *and* b. The symbol:

$$\int_a^X f(x)\,dx$$

(read "the integral of $f(x)$ from a to X") is called the integral of $f(x)$ *with an upper variable limit.*

After learning all these new shorthand symbols, which are convenient but *do not add anything* to the 3 steps of the integration process, you probably expect the process to be applied to the specific cost function $c(h)$ of our example to obtain the *exact* cost C of the building of height $H = 200$ ft, which is the integral of the function $c(h)$ from $h = 0$ to $h = H$ and by (6.4.2) is written:

$$C = \int_0^H c(h)\,dh = \int_0^{200} (60{,}000 + 1.6h^2)\,dh. \tag{6.4.4}$$

After all, differentiation by the delta method did allow us to evaluate exactly the slope at a point P_1 of any curve and you rightfully expect the integration process to do the same for the area under a curve. But, while the delta method is a general process, which allows the actual evaluation of the derivative of any function $f(x)$, *there is no corresponding general method for the evaluation of the integral of* $f(x)$. This does *not* mean that you *cannot* obtain the exact value of the cost C. It simply means that before you can, you must learn that there is a direct relationship between differentiation and integration, *which allows you to integrate without difficulty, if you only remember derivatives.* This will be shown in the next section.

Meanwhile, evaluate approximately by summation of rectangles the following integrals, obtaining approximations from above and below and using the given number n of intervals.

Note: The given values are the exact values of the integrals.

6.4.1 $\int_0^4 x\,dx = 8, \quad n = 1, 2, 4.$

6.4.2 $\int_2^8 (1 + x)\,dx = 36, \quad n = 1, 2, 3.$

6.4.3 $\int_{-1}^5 x\sqrt{1 + x^2}\,dx = 43.245, \quad n = 1, 2, 3.$

6.4.4 $\int_0^\pi \sin x\,dx = 2, \quad n = 2, 4.$

6.4.5 $\int_0^{2\pi} \cos x\,dx = 0, \quad n = 2, 4, 8.$

6.4.6 $\int_1^2 \left(\frac{1}{x^2}\right)dx = 0.5, \quad n = 2, 4, 5.$

6.4.7 $\int_0^{\pi/4} \tan x\,dx = 0.347, \quad n = 2, 4.$

6.5 THE INTEGRAL AS ANTIDERIVATIVE

You have met in arithmetic and in algebra operations that cancel each other. For example, subtraction cancels addition, since adding and subtracting the same number to and from a given number leaves the given number unchanged:

$$3 + 4 - 4 = 7 - 4 = 3.$$

Multiplication and division by the same number also cancel each other:

$$\frac{3 \times 4}{4} = \frac{12}{4} = 3.$$

"Extracting the square root" and "squaring" cancel each other:

$$(\sqrt{9})^2 = (\pm 3)^2 = 9.$$

(Notice, though, that if you *first* square and *then* take the square root, you may change the sign of the number:

$$\sqrt{(+3)^2} = \sqrt{9} = \pm 3.)$$

Two operations that cancel each other are said to be the *inverse* of each other.

You will now be shown that *differentiation and integration are, in a sense, inverse operations.*

Let us first check this assertion on two very simple functions. The integral:

$$A = \int_0^X (1)\,dx$$

is the area under the curve $f(x) = 1$ from $x = 0$ to $x = X$ (Fig. 6.7), i.e., the area of the rectangle of height 1 and width X. Hence:

$$A = \int_0^X (1)\,dx = 1 \times X = X. \qquad \text{(a)}$$

The integral of $f(x)$ is a function of the upper limit of integration X, which in general we shall indicate by $F(X)$. The integral of $f(x) = 1$ between zero and X is thus:

$$A \equiv F(X) = X. \qquad \text{(b)}$$

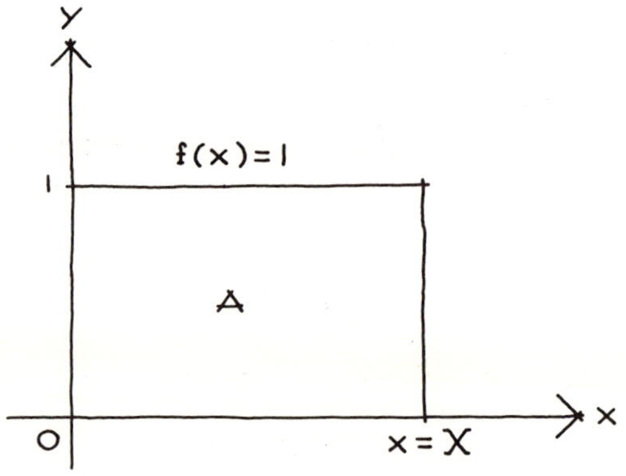

Figure 6.7

If we take the derivative of $F(X)$ with respect to its variable X, we obtain:

$$\frac{dF(X)}{dX} = \frac{dX}{dX} = 1 = f(x). \tag{c}$$

(a) and (c) show that the integral $F(X)$ between zero and X of the function $f(x) = 1$ is X, and that the derivative of the integral $F(X) = X$ is the function $f(x) = 1$. Hence, if you first integrate $f(x) = 1$ between zero and X and then differentiate the result with respect to X, you obtain back $f(x) = 1$. In this example, integration and differentiation cancel each other.

As a second example, consider the integral:

$$A \equiv F(X) = \int_0^X x\,dx, \tag{d}$$

which in Fig. 6.8 is seen to be the area of the triangle bound by the 45° line $f(x) = x$ between zero and X and hence, to be equal to $\frac{1}{2}X^2$:

$$F(X) = \int_0^X x\,dx = \frac{1}{2}X^2. \tag{e}$$

If we now differentiate $F(X) = \frac{1}{2}X^2$ with respect to X we obtain:

$$\frac{dF(X)}{dX} = \frac{d(\frac{1}{2}X^2)}{dX} = \frac{1}{2}\frac{dX^2}{dX} = \frac{1}{2}2X = X. \tag{f}$$

The derivative with respect to X of the integral of $f(x) = x$ between zero and X is X, i.e., the value of $f(x)$ at $x = X$. Since X is just any value of x, here again integration and differentiation cancel each other.

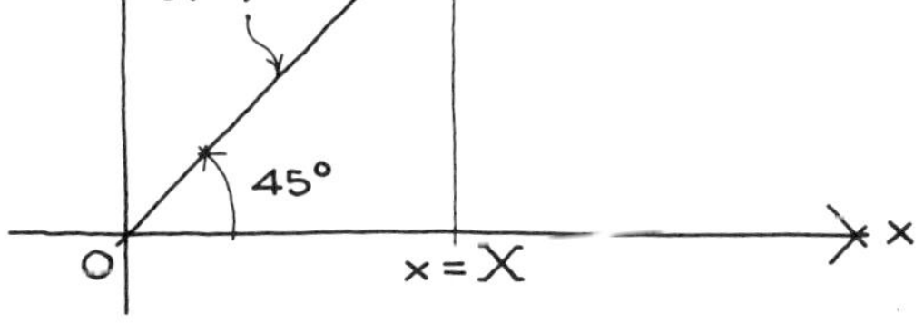

Figure 6.8

These two examples have shown you, so far, that: (1) the integral of a function $f(x)$ with an upper variable limit X is a function $F(X)$ of the upper variable limit, and (2) the derivative of $F(X)$ with respect to X is the function $f(x)$ under the integral sign *evaluated at* $x = X$. Read points (1) and (2) again and make sure you understand this statement, which, although simple, is expressed by a lot of words. This is why it is so much easier to write the statement in mathematical shorthand:

$$\text{if } F(X) = \int_0^X f(x)\,dx, \text{ then } \frac{dF(X)}{dX} = f(X). \tag{6.5.1}$$

To prove in general that the derivative of $F(X)$ is $f(X)$, all we have to do is evaluate the derivative of F with respect to X by the delta method. Here are the 5 steps of the delta method applied to $F(X)$, where $F(X)$ stands for "integrate $f(x)$ from c to X." (In the previous examples the constant c was equal to zero.)

(a) $F(X_1) = \int_c^{X_1} f(x)\,dx,$

(b) $F(X_1 + \Delta X_1) = \int_c^{X_1+\Delta X_1} f(x)\,dx,$

(c) $\Delta F_1 = F(X_1 + \Delta X_1) - F(X_1) = \int_c^{X_1+\Delta X_1} f(x)\,dx - \int_c^{X_1} f(x)\,dx.$

But, by Fig. 6.9, the first integral is the area under $f(x)$ between $x = c$ and $x = X_1 + \Delta X_1$, the second the area under $f(x)$ between $x = c$ and $x = X_1$.

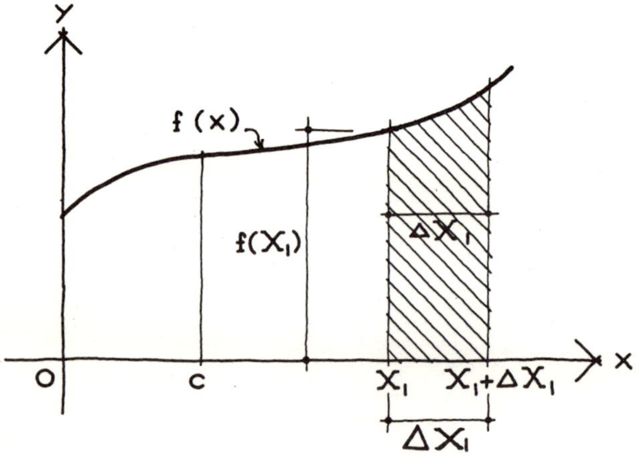

Figure 6.9

Hence, their difference is the shaded area under $f(x)$ between X_1 and $X_1 + \Delta X_1$ or:

$$\Delta F_1 = \int_{X_1}^{X_1+\Delta X_1} f(x)\,dx.$$

(d) $$\frac{\Delta F_1}{\Delta X_1} = \frac{1}{\Delta X_1}\int_{X_1}^{X_1+\Delta X_1} f(x)\,dx.$$

(e) $$\left.\frac{dF}{dX}\right]_{X=X_1} = \lim_{\Delta X_1 \to 0} \frac{1}{\Delta X_1}\int_{X_1}^{X_1+\Delta X_1} f(x)\,dx.$$

As ΔX_1 becomes smaller and smaller while X_1 stays put, the shaded area of Fig. 6.9 becomes nearer and nearer to the area of a rectangle of width ΔX_1 and height $f(X_1)$. Hence:

$$\left.\frac{dF}{dX}\right]_{X=X_1} = \frac{1}{\Delta X_1}[f(X_1)\,\Delta X_1] = f(X_1),$$

which proves our statement for any function $f(x)$.

This result is of the greatest practical importance. It proves that if we happen to know that $f(x)$ is the derivative of a function $F(x)$, we know by the same token that $F(X)$ is the integral of $f(x)$ between a certain lower limit c and an upper limit $x = X$.

For example, we know that $3x^2$ is the derivative of x^3; hence, we also know that X^3 is the integral of $3x^2$ between $x = 0$ and $x = X$ (the lower limit c equals zero in this case, since the integral, X^3, equals 0 if the upper limit $X = 0$):

$$X^3 = \int_0^X 3x^2\,dx.$$

Dividing by 3 on both sides, we obtain the integral of x^2 between zero and X:

$$F(X) = \int_0^X x^2\,dx = \frac{1}{3}X^3.$$

We must now notice that, since the derivative of a constant C is equal to zero [see (5.3.7)], all the functions $\frac{1}{3}X^3 + C$, whatever the value of the constant C, have the same derivative X^2. To put it the other way around, the functions $\frac{1}{3}X^3 + C$ are *all* integrals of x^2. They only differ because they have the value C at $X = 0$, while at $X = 0$ the integral $\frac{1}{3}X^3$ has the specific value zero. The area under $y = x^2$ *from* $x = 0$ to $x = X$ equals $\frac{1}{3}X^3$, while

$\frac{1}{3}X^3 + C$ represents the area under $y = x^2$ from some undefined value $x = c$ different from zero up to $x = X$.

The *family of functions* $\frac{1}{3}x^3 + C$ is called the *indefinite integral of* x^2 (because the lower limit of integration c and the upper limit of integration X for x are not defined), and is symbolized by the integral sign *without limits:*

$$\int x^2\, dx = \frac{1}{3}x^3 + C = F(x) + C.$$

The area under $y = x^2$ between a *given* lower limit a and a *given* upper limit b may be conveniently expressed by means of the indefinite integral of x^2, since it is the difference of the areas from c to b and from c to a (Fig. 6.10):

$$A = \int_a^b x^2\, dx = \int_c^b x^2\, dx - \int_c^a x^2\, dx = [F(b) + C] - [F(a) + C]$$
$$= F(b) - F(a) = \tfrac{1}{3}b^3 - \tfrac{1}{3}a^3.$$

The difference between $F(b)$ and $F(a)$ is symbolized by $F(x)]_a^b$ and the integral from a to b of $y = f(x)$ is written:

$$\int_a^b f(x)\, dx = F(x)\Big]_a^b \tag{6.5.2}$$

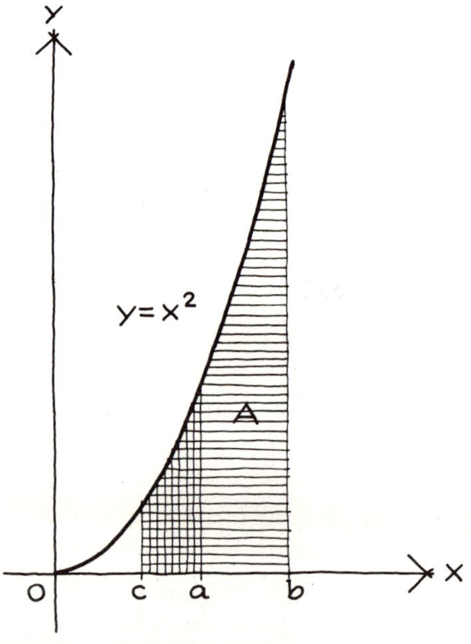

Figure 6.10

6.5.1 We can now find the exact cost C of the building 200 ft high of Section 6.4 by performing the integration of (6.4.4).

Since the area under the *sum* of the ordinates of two curves is the *sum of the area under each curve*, the integral (6.4.4) may be evaluated as the sum of two separate integrals:

$$C = \int_0^{200} (60{,}000 + 1.6\,h^2)\,dh = \int_0^{200} 60{,}000\,dh + \int_0^{200} 1.6\,h^2\,dh$$
$$= 60{,}000 \int_0^{200} (1)\,dh + 1.6 \int_0^{200} h^2\,dh.$$

(The multiplying constants 60,000 and 1.6 may be taken out of the integral sign, since when we multiply all the ordinates of a curve by a constant, the heights of all the elementary rectangles and, hence, their areas get multiplied by the same constant.)

As shown by the examples of this section, the integral from zero to X of $f(x) = 1$ is X and the integral from zero to X of $f(x) = x^2$ is $\frac{1}{3}X^3$. Hence, changing the name of the variable from x to h and with $a = 0$, $b = 200$, we obtain by (6.5.2):

$$C = [60{,}000h + 1.6(\tfrac{1}{3}h^3)]_0^{200}$$
$$= \left[60{,}000 \times 200 + \frac{1.6}{3}(200)^3\right] - \left[60{,}000 \times 0 + \frac{1.6}{3}(0)^3\right]$$
$$= 12{,}000{,}000 + \frac{1.6}{3}8{,}000{,}000 = \$16{,}266{,}667.$$

The approximate result, \$16,266,720, obtained numerically in Section 6.4 agrees to 5 significant figures with this rigorously correct result.

6.6 INDEFINITE AND DEFINITE INTEGRALS

The results of Section 6.5 show that integration requires a good memory of differentiation: reading backwards a table of derivatives one obtains a table of integrals. Thus, we know that:

$$\text{if } f(x) = \frac{1}{n+1}x^{n+1}, \text{ then } \frac{df}{dx} = x^n.$$

Hence:

$$\text{if } f(x) = x^n, \text{ then } \int x^n\,dx = \frac{1}{n+1}x^{n+1} + C \qquad (n \neq -1). \qquad (6.6.1)$$

Similarly, we know that:

$$\text{if } f(x) = \sin x, \text{ then } \frac{df}{dx} = \cos x,$$

and consequently that:

$$\text{if } f(x) = \cos x, \text{ then } \int \cos x\, dx = \sin x + C. \qquad (6.6.2)$$

We know that:

$$\text{if } f(x) = \cos x, \text{ then } \frac{df}{dx} = -\sin x;$$

hence:

$$\text{if } f(x) = -\sin x, \text{ then } \int -\sin x\, dx = \cos x + C,$$

or:

$$\text{if } f(x) = \sin x, \text{ then } \int \sin x\, dx = -\cos x + C. \qquad (6.6.3)$$

But since we do *not* yet know of any function whose derivative is tan x, so far the integral of tan x is *unknown* to us. Many integrals of other *known* functions are also *unknown* for the same reason. This may be frustrating, but is unavoidable, since there is no general process to integrate a given function. For example, even (6.6.1) fails in one particular case, the case $f(x) = 1/x = x^{-1}$, since it gives:

$$\int \frac{dx}{x} = \frac{1}{-1+1} x^{-1+1} + C = \frac{x^0}{0} + C.$$

But by definition $x^0 \equiv 1$:

$$\int \frac{dx}{x} = \frac{1}{0} + C = \infty + C = \infty.$$

(This difficulty is due to the fact that the curve $f(x) = 1/x$ grows forever as $x \longrightarrow 0$ so that the "first rectangle" $(1/x)\, dx$ of the summation has an infinite height and zero width, i.e., an incomputable area.) To find the integral of $1/x$ one must actually *invent* a new function, whose derivative equals $1/x$. Such a function has been invented and is tabulated in Table 3 of the Appendix. It is called *the natural log of* x and is symbolized by:

$$f(x) = \ln x \quad \text{(read "natural log of } x\text{")}.$$

Then, by definition:

$$\frac{d \ln x}{dx} = \frac{1}{x} \qquad \therefore \quad \int \frac{1}{x}\, dx = \ln x + C. \qquad (6.6.4)$$

The integral of $\tan x = \sin x/\cos x$ may now be obtained by noticing that:

$$\frac{d \cos x}{dx} = -\sin x \qquad \therefore \quad \frac{d \cos x}{dx}\, dx = d \cos x = -\sin x\, dx,$$

so that:

$$\int \tan x\, dx = \int \frac{\sin x\, dx}{\cos x} = -\int \frac{d \cos x}{\cos x}.$$

Letting $\cos x = z$ and using (6.6.4):

$$-\int \frac{d \cos x}{\cos x} = -\int \frac{1}{z}\, dz = -\ln z + C = -\ln(\cos x) + C.$$

(To evaluate $\ln(\cos x)$ for a given x you first find the value z of $\cos x$ in Table 2 and then look in Table 3 for the value of $\ln x$ for $x = z$.) Thus:

$$\int \tan x\, dx = -\ln(\cos x) + C. \tag{6.6.5}$$

In view of the essential difficulty in obtaining integrals, most of us who use mathematics as a practical tool simply refer to a table of integrals to find the indefinite integral of a given function. One such brief table appears in the Appendix (page 158).

In conclusion, integration, while possibly difficult in theory, is simple in practice: it is performed mostly by looking in a table.

In any case, just to make sure you know how to use a table of integrals, try to get the following integrals by yourself, through your memory of derivatives.

6.6.1 $\int x^4\, dx = \frac{1}{5}x^5 + C.$

6.6.2 $\int \sqrt{x}\, dx = \frac{2}{3}\sqrt{x^3} + C.$

6.6.3 $\int \sin 2x\, dx = -\frac{1}{2}\cos 2x + C.$

Hint: Let $2x = z$ and notice that $dz = (dz/dx)\, dx = 2\, dx$, so that $dx = \frac{1}{2}\, dz$.

6.6.4 $\int \frac{dx}{\sqrt{x}} = 2\sqrt{x} + C.$

6.6.5 $\int \frac{\cos x\, dx}{\sin x} = \ln(\sin x) + C.$

Hint: Let $\sin x = z$.

6.6.6 $\int \sin x \cos x \, dx = \frac{1}{2} \sin^2 x + C.$

Hint: Let $\sin x = z$.

In your work, you will use integration mostly to evaluate definite integrals, i.e., integrals with specific upper and lower limits. As noticed in Section 6.5, this is simply done by (6.5.2), i.e., by taking the difference between the values of the indefinite integral at the upper and at the lower limits of integration. For example:

$$\int_2^5 x^3 \, dx = \frac{1}{4} x^4 \Big]_2^5 = \frac{1}{4}(5^4 - 2^4) = \frac{1}{4}(625 - 16) = 152.25.$$

You may now evaluate the following definite integrals, that is, the areas under the given functions between the given limits.

6.6.7 $\int_1^4 x^2 \, dx = 21.$

6.6.8 $\int_{-1}^3 x^4 \, dx = 48.8.$

6.6.9 $\int_{-\pi}^{\pi} \cos x \, dx = 0.$

6.6.10 $\int_{\pi/4}^{3\pi/4} \sin x \, dx = \sqrt{2} = 1.414.$

6.6.11 $\int_2^4 \frac{dx}{\sqrt{x}} = 2(\sqrt{4} - \sqrt{2}) = 1.172.$

6.6.12 $\int_0^{\pi/2} \sin 2x \, dx = 1.$

6.6.13 $\int_0^{\pi/4} \tan 3x \, dx = -\frac{1}{3}\left[\ln\left(\cos \frac{\pi}{4}\right) - \ln(\cos 0)\right] = -\frac{1}{3} \ln 0.707 = 0.116.$

6.7 APPLICATIONS OF INTEGRATION

(a) *Centroids*

We call the *centroid* of a plane area or a plane curve (Fig. 6.11a, 6.11b) the point C on which one can balance on a pin the figure or the curve, if these are considered to be of uniform weight per unit area or length. The point C is the intersection of two lines (or knife edges) on which one

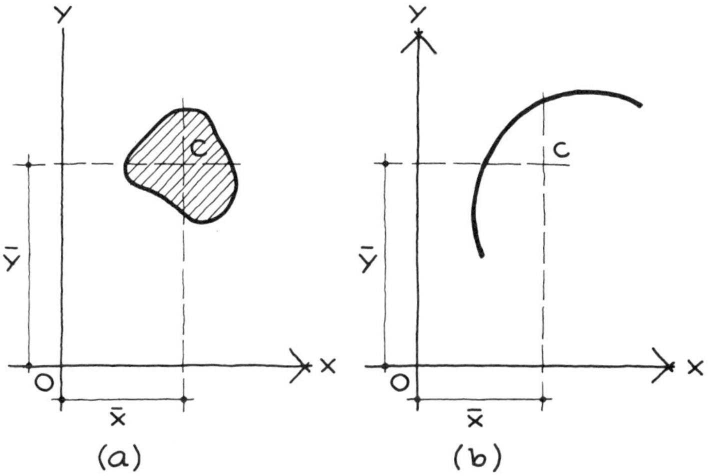

Figure 6.11

can balance the figure or curve. The centroid plays an important role in all structural problems.

When the plane figure or curve has an axis of symmetry, it is balanced on this axis; hence, the centroid must lie on it (Fig. 6.12a). When the figure or line has two axes of symmetry, the centroid must lie on both and, hence, is at the intersection of the axes (Fig. 6.12b).

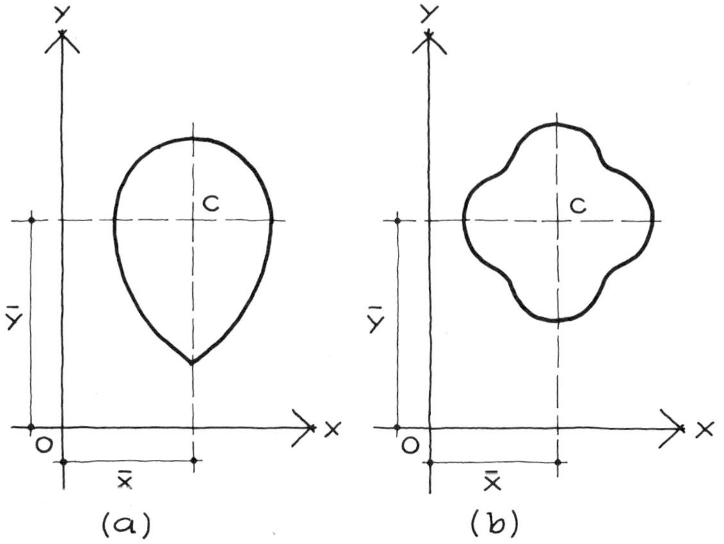

Figure 6.12

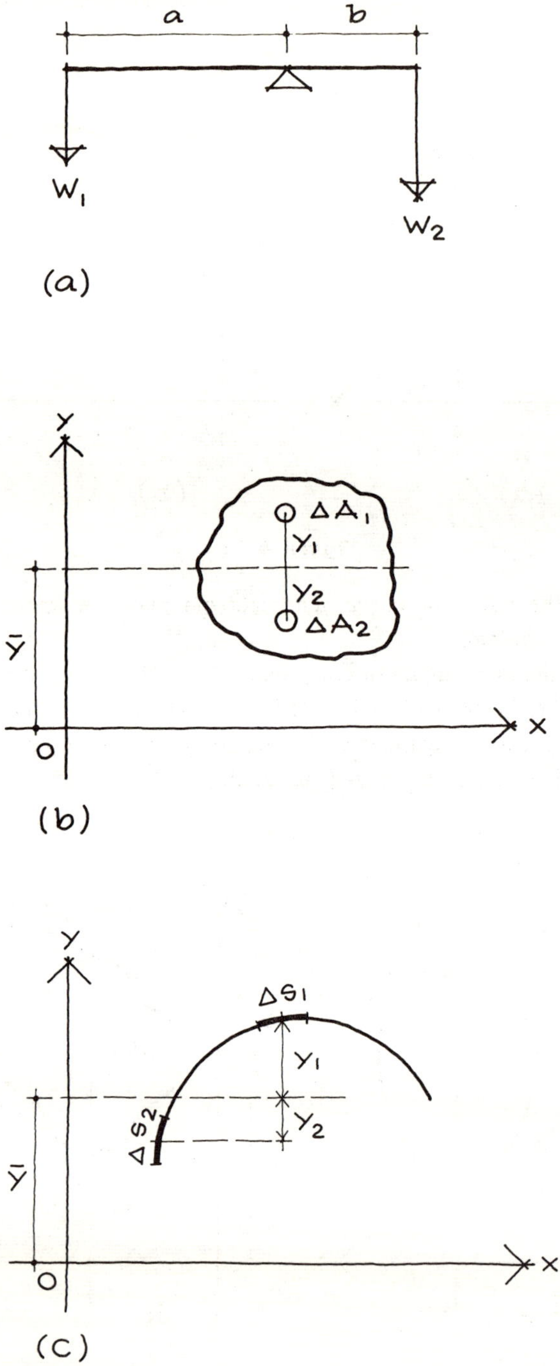

Figure 6.13

By the principle of the lever, two weights W_1, W_2 connected by a bar are in balance over a knife edge (Fig. 6.13a) if the products of the weights times their distance from the knife edge are equal. Hence, for the figure (or line) to be balanced on an axis, the lever principle must be satisfied by the sum of the products of all the elementary areas ΔA (or lengths Δs) into which the figure or line may be decomposed, times the corresponding distances (Fig. 6.13b, c):

$$\sum y_{i,1}\,\Delta A_{i,1} = \sum y_{i,2}\,\Delta A_{i,2}; \qquad \sum y_{i,1}\,\Delta s_{i,1} = \sum y_{i,2}\,\Delta s_{i,2},$$

or, bringing the sum on the right to the left side of the equation and indicating distances *below* the axis as negative and above as positive:

$$\sum y_i\,\Delta A_i = 0 \qquad \text{or} \qquad \sum y_i\,\Delta s_i = 0. \tag{6.7.1}$$

6.7.1 Locate the centroid of the perimeter of a half-circle (Fig. 6.14).

Since the circular curve ABO is symmetrical about an axis parallel to y through B, the curve (you may think of it as a thin wire) must only be balanced on a knife edge parallel to x through a point C. Let us consider an element of the circular curve of length $rd\theta$. Its midpoint has a distance $y = r\sin\theta$ from the x-axis and its distance from the centroidal knife edge DE is $y - \bar{y} = r\sin\theta - \bar{y}$. By (6.7.1) the sum of all the lengths $rd\theta$ times the corresponding distance $y - \bar{y}$ must add up to zero for balance. The sum is actually an integral since the elements $rd\theta$ are infinitesimal; hence, varying θ from 0 to π to take into account all the elements of the half-circle:

$$\int_0^\pi (r\sin\theta - \bar{y})rd\theta = 0$$

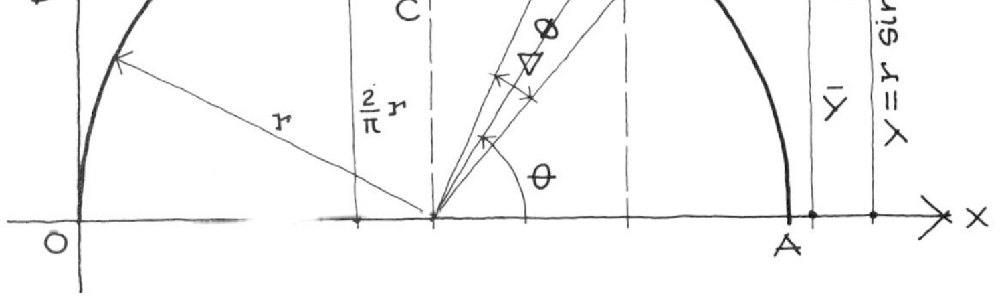

Figure 6.14

or:

$$r^2 \int_0^\pi \sin\theta \, d\theta - \bar{y} r \int_0^\pi d\theta = r^2[-\cos\theta]_0^\pi - \bar{y} r \theta]_0^\pi$$
$$= r^2[-\cos\pi - (-\cos 0)] - \bar{y} r(\pi - 0)$$
$$= 2r^2 - \bar{y}\pi r = 0 \quad \therefore \quad \bar{y} = \frac{2}{\pi} r.$$

6.7.2 Locate the centroid of the "T-section" in Fig. 6.15, taking as elementary areas rectangles of height dy and width equal to the width of the section at y.

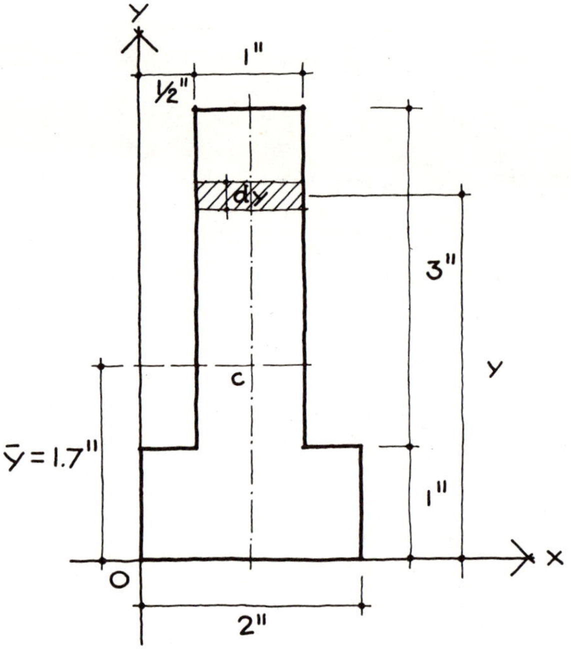

Figure 6.15

$$\int_1^4 (y - \bar{y})(1 \times dy) + \int_0^1 (y - \bar{y})(2 \times dy) = 0. \qquad \left(\bar{y} = \frac{17}{10} = 1.7 \text{ in.}\right)$$

6.7.3 Locate the centroid of the "angle-section" of Fig. 6.16.

The 45° line through O, $y = x$, is a line of symmetry on which C must lie. To determine $\bar{y}(= \bar{x})$ by (6.7.1):

$$\int_{1/2}^5 (y - \bar{y})\left(\frac{1}{2} \times dy\right) + \int_0^{1/2} (y - \bar{y})(5 \times dy) = 0. \qquad \left(\bar{y} = \frac{109}{76} = 1.43 \text{ in.}\right)$$

6.7.4 Evaluate the distance BC between the centroid C of the perimeter of a quarter of a circle and the point B of the circle on a 45° line through its center O (Fig. 6.17).

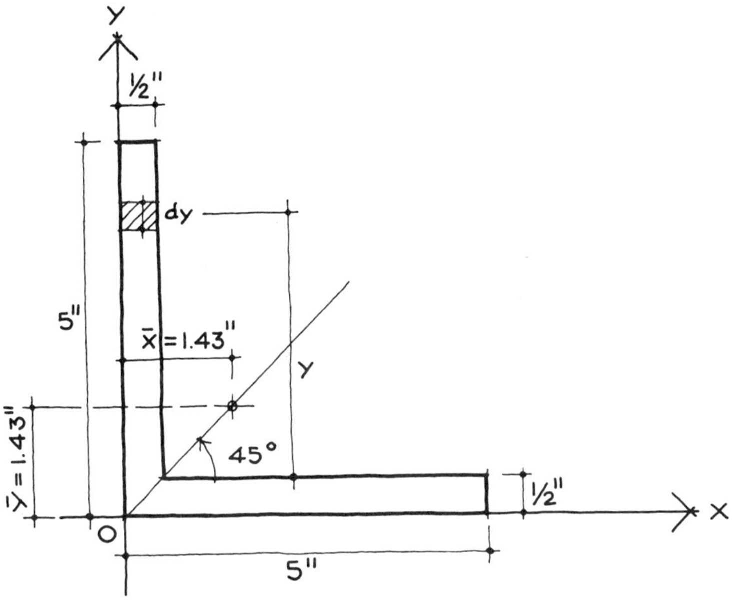

Figure 6.16

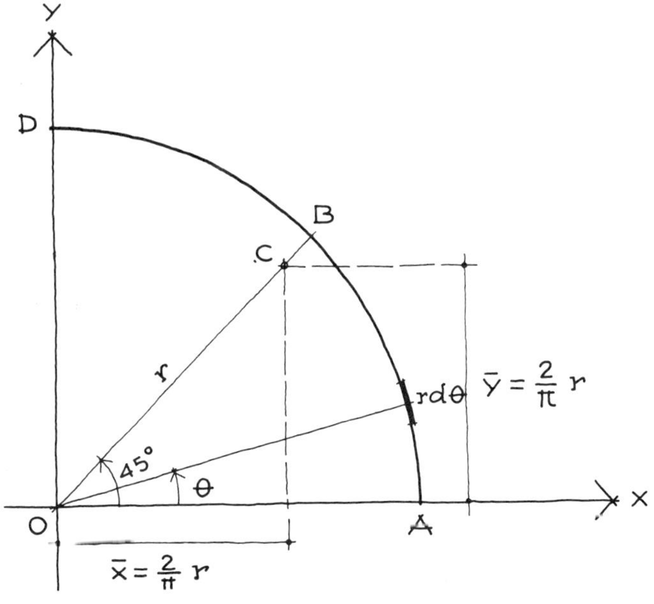

Figure 6.17

The 45° line through O is a line of symmetry for the arc ABD. To determine $\bar{y}(= \bar{x})$ by (6.7.1):

$$\int_0^{\pi/2} (r \sin \theta - \bar{y})\, r\, d\theta = 0.$$

$$\left(BC = \left(1 - \frac{2\sqrt{2}}{\pi}\right) r = 0.10r.\right)$$

(b) *Moments of inertia*

The rigidity of a beam is inversely proportional to its length and proportional to the product of a constant that depends on the material times another constant that depends on the shape of its cross section. This last constant is called the *moment of inertia* of the cross section and for sections with an axis of symmetry it is the sum of the elementary areas multiplied by the *square* of their distance from an axis through the centroid perpendicular to the axis of symmetry.

6.7.5 Evaluate the moment of inertia I of a rectangle $b \times h$ with respect to an axis parallel to the x-axis through its centroid (Fig. 6.18).

$$I = \int_0^h \left(y - \frac{h}{2}\right)^2 b\, dy = b \int_0^h \left(y - \frac{h}{2}\right)^2 d\left(y - \frac{h}{2}\right).$$

(Note that $dy = d(y - h/2)$ since the derivative of a constant is zero:

$$d\left(y - \frac{h}{2}\right) = \frac{d(y - h/2)}{dx}\, dx = \frac{dy}{dx}\, dx - \frac{d(h/2)}{dx}\, dx = \frac{dy}{dx}\, dx = dy.)$$

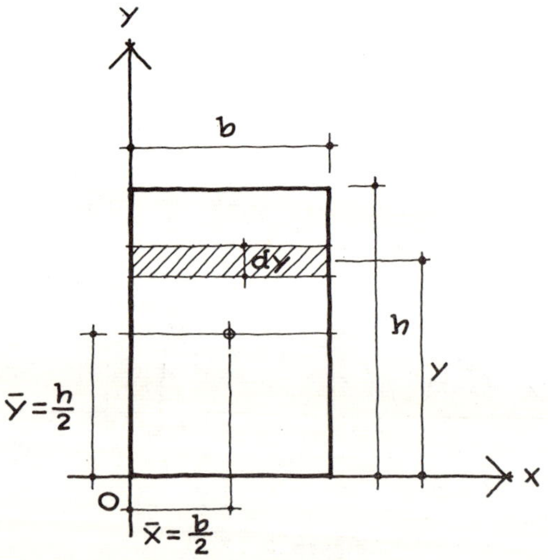

Figure 6.18

Let $y - h/2 = z$:

$$\int \left(y - \frac{h}{2}\right)^2 d\left(y - \frac{h}{2}\right) = \int z^2\, dz = \frac{1}{3}z^3 + C = \frac{1}{3}\left(y - \frac{h}{2}\right)^3 + C,$$

$$b\int_0^h \left(y - \frac{h}{2}\right)^2 d\left(y - \frac{h}{2}\right) = \frac{b}{3}\left(y - \frac{h}{2}\right)^3\Big]_0^h = \frac{b}{3}\left[\left(h - \frac{h}{2}\right)^3 - \left(0 - \frac{h}{3}\right)^3\right]$$

$$= \frac{b}{3}\left(\frac{h^3}{8} + \frac{h^3}{8}\right) = \frac{1}{12}bh^3.$$

For $b = 1.8$ in., and $h = 5$ in., $I = \frac{1}{12}(1.8)(5)^3 = 18.75$ in.^{4}

6.7.6 Evaluate the moment of inertia of the symmetrical I-section of Fig. 6.19 about the centroidal x-axis.

Notice that the contribution to the moment of inertia of the lower half of the section is equal to the contribution of its upper half.

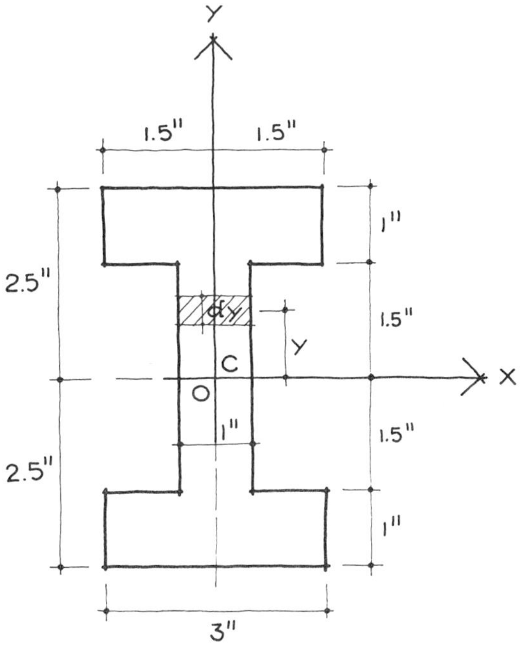

Figure 6.19

$$I = 2\left\{\int_0^{1.5} y^2\,(1 \times dy) + \int_{1.5}^{2.5} y^2\,(3 \times dy)\right\}. \qquad (I = 26.75 \text{ in.}^4)$$

Note: The I-section of this problem has the same area and the same depth as the rectangular section of problem 6.7.5, but its moment of inertia is greater. I-sections have been developed in order to obtain greater rigidity without increase in material.

6.8 THE NATURAL GROWTH FUNCTION

We wish to end this chapter (and this book) by introducing you to an interesting and useful function, called the *exponential function*, which is *equal to its derivative at every point* and is strictly related to the natural log function.

Consider putting \$1.00 in a bank at 100% interest per year. After a year you will have in the bank $1 + (\frac{100}{100})(1) = (1 + 1) =$ \$2.00. Let us now assume that your bank compounds interest every 6 months so that, at 100% interest *per year*, i.e., at 50% interest for half a year, you will have after the first 6 months $1 + (\frac{50}{100})(1) = (1 + \frac{1}{2})(1) =$ \$1.50. During the second 6 months you get, in addition, 50% on \$1.50 or $(\frac{50}{100})$(\$1.50) = \$0.75, so that at the end of a year you will have

$$1.50 + (\tfrac{50}{100})(1.50) = (1.50)(1 + 0.50) = (1 + 0.50)^2$$
$$= (1 + \tfrac{1}{2})^2 = \$2.25.$$

If interest is compounded every 4 months, i.e., every third of a year, you will have $(1 + \frac{1}{3})^3 =$ \$2.37 at the end of a year. If the interest is compounded more and more often, the money at the end of a year grows, but, as shown by Table 6.3, *not* indefinitely. At most you will get \$2.718 . . . if the interest

Table 6.3

n	1	2	3	4	5	10	20	100	1,000	10,000
$\left(1 + \frac{1}{n}\right)^n$	2	2.25	2.37	2.44	2.49	2.59	2.65	2.70	2.71	2.718

were compounded "at every instant." This limit number occurs so often in mathematical calculations that it has a name (just as 3.14159 . . . has the name π). It is called the *base of Napierian logs* and is symbolized by the letter e:

$$e = \lim_{n\to\infty}\left(1 + \frac{1}{n}\right)^n = 2.718281\ldots. \tag{6.8.1}$$

Consider now the function obtained by raising the number e to the power x:

$$y = e^x, \tag{6.8.2}$$

which is called the *exponential function* and graphed versus x in Fig. 6.20. Let us prove by the delta method that the derivative of $y = e^x$ is equal to e^x.

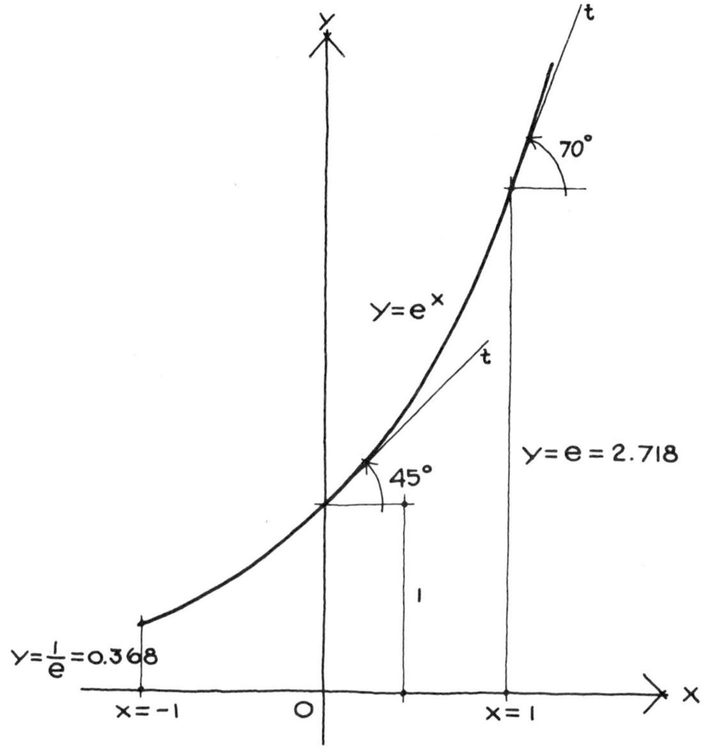

Figure 6.20

(a) $f(x_1) = e^{x_1}$,

(b) $f(x_1 + \Delta x_1) = e^{x_1+\Delta x_1} = e^{x_1}\, e^{\Delta x_1}$ [see (2.3.1)],

(c) $\Delta y_1 = f(x_1 + \Delta x_1) - f(x_1) = e^{x_1}e^{\Delta x_1} - e^{x_1} = e^{x_1}(e^{\Delta x_1} - 1)$,

(d) $\dfrac{\Delta y_1}{\Delta x_1} = e^{x_1}\dfrac{e^{\Delta x_1} - 1}{\Delta_{x_1}}$,

(e) $\left.\dfrac{dy}{dx}\right]_{x=x_1} = \lim\limits_{\Delta x_1 \to 0} \dfrac{\Delta y_1}{\Delta x_1} = e^{x_1} \lim\limits_{\Delta x_1 \to 0} \dfrac{e^{\Delta x_1} - 1}{\Delta x_1}$,

since x_1, and hence e^{x_1} remain unchanged as Δx_1 approaches zero.

To evaluate this limit numerically in an easy manner, let Δx_1 approach zero by taking the values $\Delta x_1 = 1, \frac{1}{2}, \frac{1}{4}, \frac{1}{8}, \ldots$, so that $e^{\Delta x_1}$ takes the values $e^1 = e = 2.718$, $e^{1/2} = \sqrt{e} = \sqrt{2.718} = 1.649$, $e^{1/4} = \sqrt{e^{1/2}} = \sqrt{1.649} = 1.285, \ldots$, obtainable by successive square roots. Table 6.4 gives the corresponding values of $e^{\Delta x_1}$ and of the ratio $(e^{\Delta x_1} - 1)/\Delta x_1$.

Table 6.4

Δx_1	1	$\frac{1}{2}$	$\frac{1}{4}$	$\frac{1}{8}$	$\frac{1}{16}$	$\frac{1}{32}$
$e^{\Delta x_1}$	2.718	1.649	1.284	1.133	1.064	1.032
$(e^{\Delta x_1} - 1)/\Delta x_1$	1.718	1.297	1.136	1.065	1.032	1.016
Δx_1	$\frac{1}{64}$	$\frac{1}{128}$	$\frac{1}{256}$	$\frac{1}{512}$	$\frac{1}{1024}$	
$e^{\Delta x_1}$	1.016	1.008	1.004	1.002	1.00096	
$(e^{\Delta x_1} - 1)/\Delta x_1$	1.008	1.004	1.002	1.001	1.00045	

The trend in Table 6.4 shows clearly that this ratio approaches 1 as $\Delta x_1 \longrightarrow 0$. Hence:

$$\frac{de^x}{dx} = e^x, \tag{6.8.3}$$

as stated above.

For example, at $x = 0$ the slope of $y = e^x$ is $e^0 = 1$, i.e., the curve makes 45° with the x-axis; at $x = 2$, the slope is $e^2 = 7.40$; at $x = -1$, the slope is $e^{-1} = 1/e = 0.368$.

Since e^x is the derivative of e^x, by the definition of integral as antiderivative:

$$\int e^x \, dx = e^x + C. \tag{6.8.4}$$

Table 4 of the Appendix gives the values of e^x for x between -10 and $+10$. By the law of exponents the value of e^x for any other value of x is obtained by multiplication of the values in the table. For example:

$$e^{3.55} = e^{3.5} \times e^{0.05} = 33.12 \times 1.051 = 34.81.$$

The exponential function is also called the *natural growth function* because its graph is the graph of any quantity that *grows at a rate proportional to the quantity* itself. For example, the rate of change of a population at a given time is *proportional* to the number of people alive at that time. Indicating the population to be y at a time x, the law of population growth is stated mathematically by equating the rate of change of y, that is its derivative dy/dx, to ky, where k is a constant:

$$\frac{dy}{dx} = ky. \tag{6.8.5}$$

For $k = 1$, (6.8.5) is satisfied by $y = e^x$, as shown by (6.8.3). For any other value of k, the solution of (6.8.5) is given by:

$$y = e^{kx} \tag{6.8.6}$$

since, differentiating y with respect to x by the chain rule, one finds:

$$z = kx, \qquad y = e^{kx} = e^z,$$

$$\frac{dz}{dx} = k, \qquad \frac{dy}{dz} = e^z,$$

$$\frac{dy}{dx} = \frac{dy}{dz}\frac{dz}{dx} = e^z \times k = ke^{kx} = ky.$$

Evaluate e^x and e^{kx} at the given values of x by Table 4 of the Appendix.

6.8.1 $e^{3.42} = 30.57$

6.8.2 $e^{-4.45} = 0.0117$

6.8.3 $e^{10.6} \doteq 40{,}134$

6.8.4 $e^{-3.23} = 0.0396$

6.8.5 $e^{0.02x}$ for $x = 1, 2, 4, 8, 16$
($y = 1.0202, 1.0408, 1.0833, 1.1735, 1.3771$.)

6.8.6 $e^{-0.1x}$ for $x = -2, -1, 0, 1, 2$
($y = 1.2214, 1.1052, 1, 0.9048, 0.8187$.)

6.8.7 The population of the United States was 200,000,000 people in 1967. It grows 2% per year. What will it be in the year 2000?
In 33 years at $2\% = 0.02$ the population will become:

$$\begin{aligned} P &= 200{,}000{,}000\ e^{kx} = 200{,}000{,}000\ e^{0.02\times 33} \\ &= 200{,}000{,}000\ e^{0.66} = 200{,}000{,}000 \times 1.935 \\ &= 387{,}000.000. \end{aligned}$$

6.8.8 The population of New York City was 14,000,000 people in 1967. Its death rate was 2%, its birth rate was 3.5%. What will the population be in the year 2000 if these rates do not change? ($P = 22{,}967{,}000$).

The natural log function you met in Section 6.5 is nothing else but the *inverse* of the exponential function e^x, since it is the exponent to which one must raise the number e to obtain a given number. In other words, if:

$$y = e^x,$$

then

$$x = \ln y.$$

You may check that this is true by noticing that if you change the role of x and y in Table 4 for the exponential function you obtain Table 3 for the natural log function. For example, in Table 3 for $x = 0.607$, $y = -0.499 \simeq -0.50$, while in Table 4 for $x = -0.50$, $y = 0.607$.

The combination of exponentials:

$$\sinh x = \frac{e^x - e^{-x}}{2}, \qquad (6.8.7)$$

$$\cosh x = \frac{e^x + e^{-x}}{2}, \qquad (6.8.8)$$

called the *hyperbolic sine* and the *hyperbolic cosine* (read "sinch x" and "cosh x") also occur often in mathematical calculations and are graphed in Fig. 6.21a and 6.21b. The graph of the cosh x function (Fig. 6.21b) is the shape of a cable or chain hanging under its own weight and called a *catenary*.

Evaluate sinh x and cosh x at the given values of x by means of Table 4 of the Appendix.

6.8.9 $\sinh 0 = 0.$

6.8.10 $\cosh 0 = 1.$

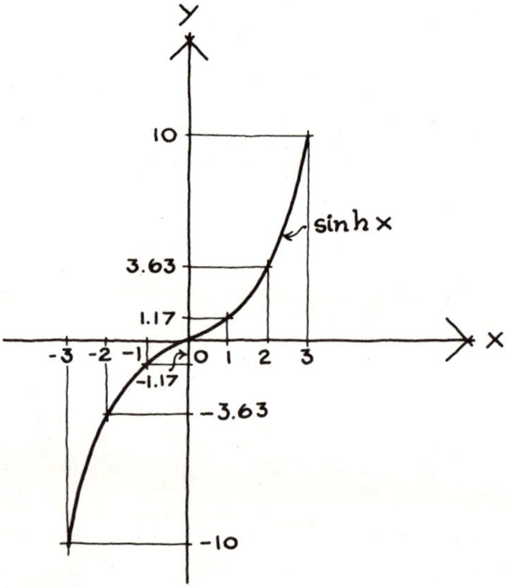

Figure 6.21(a)

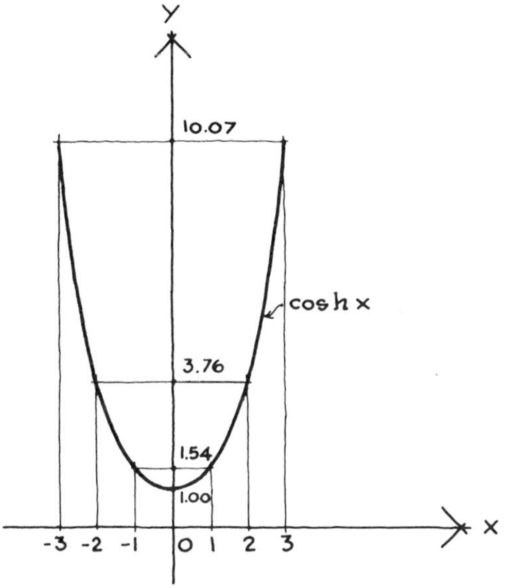

Figure 6.21(b)

6.8.11 sinh 0.5 = 0.521.

6.8.12 cosh 0.5 = 1.128.

6.8.13 cosh (−0.5) = 1.128.

6.8.14 sinh (−0.5) = −0.521.

6.8.15 Prove that $\cosh^2 x - \sinh^2 x = 1$.

The hyperbolic sine and cosine have certain analogies with the sine and cosine functions. For example, compare 6.8.9 and 6.8.10 with:

$$\sin 0 = 0, \qquad \cos 0 = 1,$$

6.8.11 and 6.8.14 with:

$$\sin(-x) = -\sin x,$$

6.8.12 and 6.8.13 with:

$$\cos(-x) = \cos x,$$

and 6.8.15 with:

$$\cos^2 x + \sin^2 x = 1.$$

Similarly, the hyperbolic tangent is defined by:

$$\tanh x = \frac{\sinh x}{\cosh x} = \frac{e^x - e^{-x}}{e^x + e^{-x}} \tag{6.8.9}$$

and is analogous to:

$$\tan x = \frac{\sin x}{\cos x}.$$

tanh x is graphed in Fig. 6.21c.

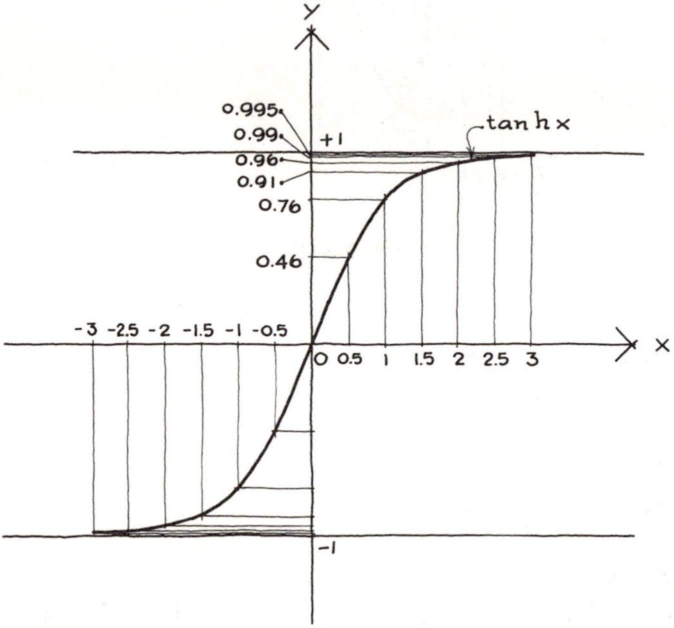

Figure 6.21(c)

6.8.16 Evaluate by means of Table 4: tanh x for $x = 1, 2, 3, 4$.
(tan $x = 0.762, 0.964, 0.995, 0.999$).

Note: As x approaches infinity, tanh x approaches 1.

6.8.17 What is the limit of tanh x as x approaches minus infinity?

$$(\lim_{x \to -\infty} \tanh x = -1).$$

CONCLUSIONS

If you were interested enough in mathematics to read this book carefully and to solve most, if not all, of its problems, you must have realized by now that mathematics is *not* beyond you and that it is of practical value.

On the other hand, you should also have noticed that even the simplest book on mathematics cannot be read as a newspaper, at a glance, because the shorthand of mathematics is so concise that reading one line of mathematics is equivalent to reading ten or one hundred lines of English. Hence, one page of this book may have taken you a good time of careful reading, even if you usually are a "fast reader." (In mathematics, actually, a "slow reader" is faster than a fast reader.)

The real usefulness of mathematics cannot be appreciated until it is applied regularly to the solution of one's problems. One of the purposes of this elementary treatment was to give you enough confidence in your mathematical ability to have you start along this road. While you should already be able to solve most of the simple prolems of your daily practice, you may encounter later on a few problems requiring more sophisticated mathematical techniques. In our technological era more and more problems are tackled scientifically, which were solved "intuitively" or "by trial and error" until a short time ago. If and when it will become profitable for you to improve your mathematical knowledge, you will find that what you have just learned is of basic importance. *But be ready then to discover much, much wider horizons.*

Meanwhile, good work.

APPENDIX

A.1 THE SLIDE RULE

You cannot solve practical mathematical problems without a slide rule, but it takes only 2 to 3 hours to become familiar with its use. By learning how to perform elementary operations on the slide rule, you will become much more confident in your work and a different person in the eye of the engineer. Later on, you may even find it fun to become a "slide-rule virtuoso," but you do not have to now.

Buy yourself a 10-in. inexpensive slide rule with not too many scales. Do *not* start to read the pamphlet enclosed in the box, but read first this section with the slide rule in front of you. Then read the pamphlet, if you like.

The slide rule (Fig. A1) consists of 3 parts:

(a) The body,
(b) The slide,
(c) The indicator with the hairline.

For the time being, notice only the two *identical* scales, one directly

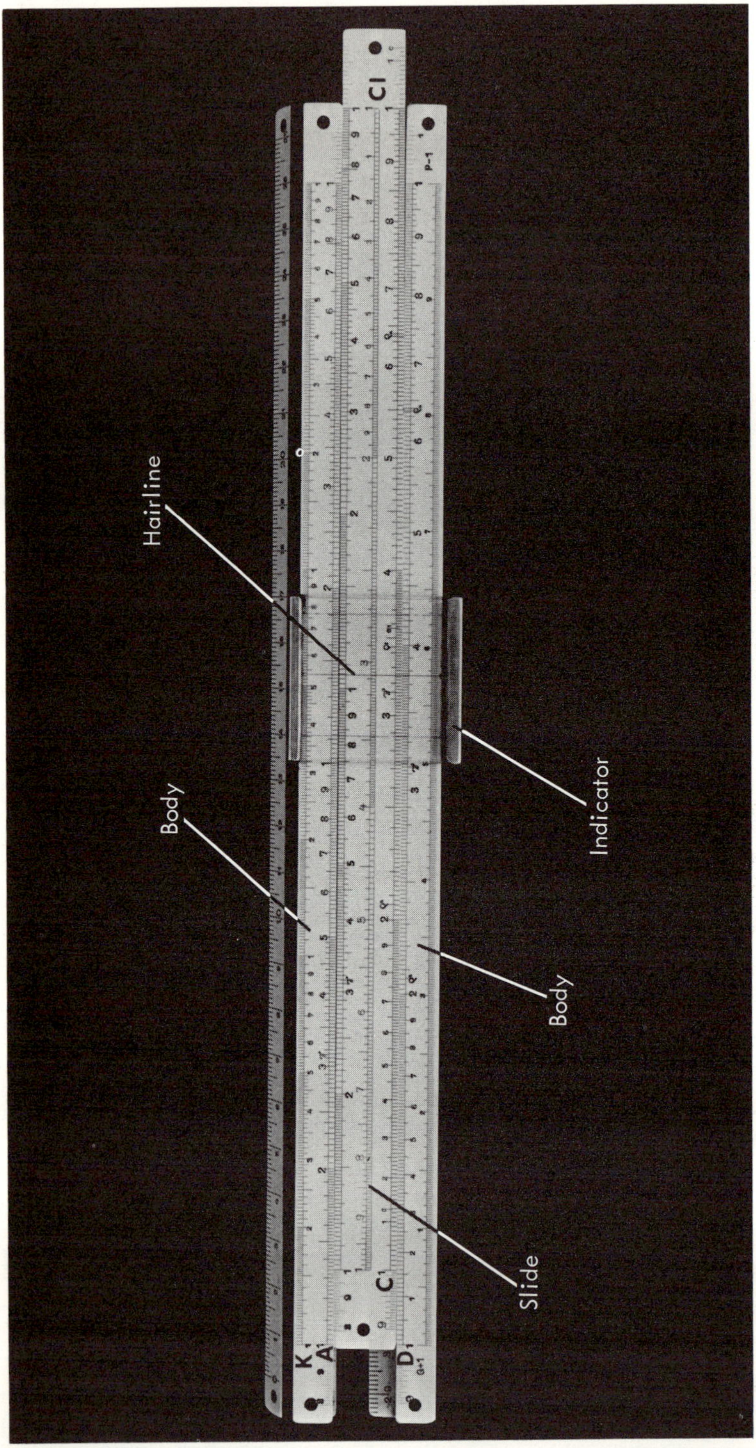

Figure A1

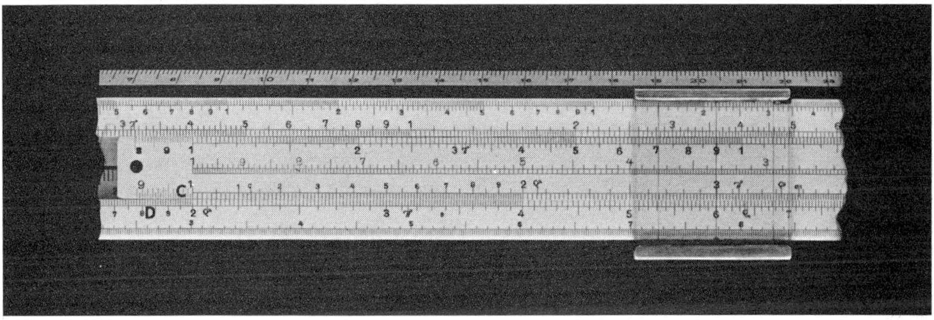

Figure A2

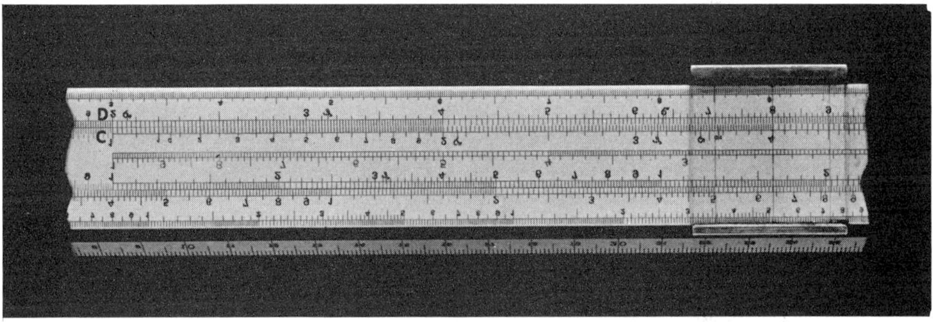

Figure A3

above the other, on the lower part of the body (scale D)* and the lower part of the slide (scale C), marked with large numbers 1, 2, 3, 4, 5, 6, 7, 8, 9 and 1 (Fig. A1). Set the left end of the slide scale C, marked big 1, over the big 2 on the body scale D, by sliding the slide to the right; now slide the hairline to the big 3 on the slide scale C (Fig. A2). You will find that the hairline falls on the 6 of the body scale D. *You have just learned to multiply* $2 \times 3 = 6$.

Do not touch the slide, but slide the hairline to the 4 on the slide scale C (Fig. A3). As you probably expect, the hairline falls on 8 on the body scale D, since $2 \times 4 = 8$. Slide the hairline to 5 on the slide scale C and you will find that the right end of the body scale D, marked 1, falls under the hairline (Fig. A4). In this case the right end 1 stands for 10 and the slide rule tells you that $2 \times 5 = 10$, but it does *not* tell you where the decimal point falls.

Similarly, $20 \times 30 = 600$ and $0.2 \times 0.3 = 0.06$, but from the position

*The scales of the slide rule are labeled with different letters by different manufacturers. Check your slide rule before reading this section.

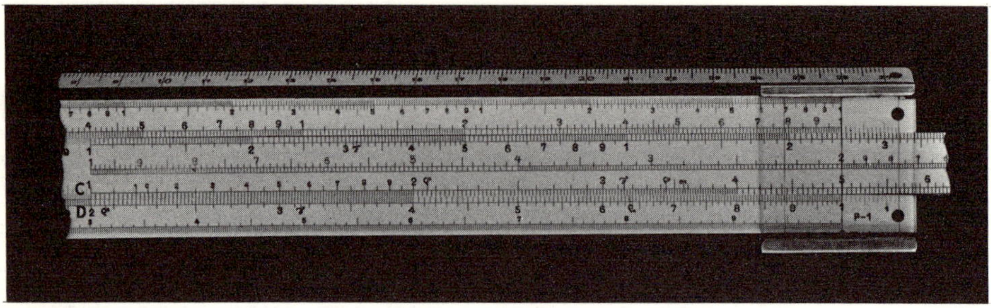

Figure A4

of the slide in Fig. A2 you can only find that $2 \times 3 = 6$. *In order to locate the decimal point you must use the "powers of* 10" *rule* of Section 1.2.

$$2 \times 3 = 6; 20 \times 30 = 2 \times 10 \times 3 \times 10 = 6 \times 10^2 = 600;$$

$$0.2 \times 0.3 = 2 \times 10^{-1} \times 3 \times 10^{-1} = 6 \times 10^{-2} = 0.06.$$

You cannot multiply 2×6 in the present position of the slide, because the 6 on the slide scale C falls outside the body scale D. But before learning how to obtain 2×6, you may want to notice that, if in the present position of the slide you have the hairline on the 3 of the slide and, hence, on the 6 of the body (Fig. A2), *you are ready to divide* 6 *by* 3. You read 6 on the body scale D, 3 on top of it on the slide scale C and find the quotient 2 on the body scale D below the left end 1 of the slide.

Now shift the hairline to the 4 of the body scale D; the 2 of the slide is on top of it and the quotient is again 2 on the body scale D, under the left 1 of the slide (Fig. A3). You have obtained $4 \div 2 = 2$. Essentially, you have already learned how to multiply and divide on the slide rule. Try other multiplications and divisions of integers, using 3 or 4 as we used 2 above.

Now let us obtain 2×6. Since this multiplication could not be performed by setting the *left* end 1 of the slide scale C on the 2 of the body scale D, set the *right* end 1 of the slide scale C on the 2 of the body scale D (Fig. A5), shift the hairline to the 6 of the slide (just as you did before with 3, 4 and 5) and you will find on the body scale D under the hairline a *small* 2 between the left big 1 and the big 2. You read this "big 1 small 2" as 12, since $2 \times 6 = 12$. Slide the hairline to the 7 on the slide scale C (Fig. A6); you read below on the body scale D "big 1 small 4," that is, 14. Try now the following operations.

1. $3 \times 2 = 6$. Set left 1 of slide on 3 of body.
2. $11 \times 5 = 55$. Set left 1 of slide on 11 (big 1 small 1) of body (Fig. A7). Set hairline on 5 of slide and read on body a mark halfway between 5 and 6.

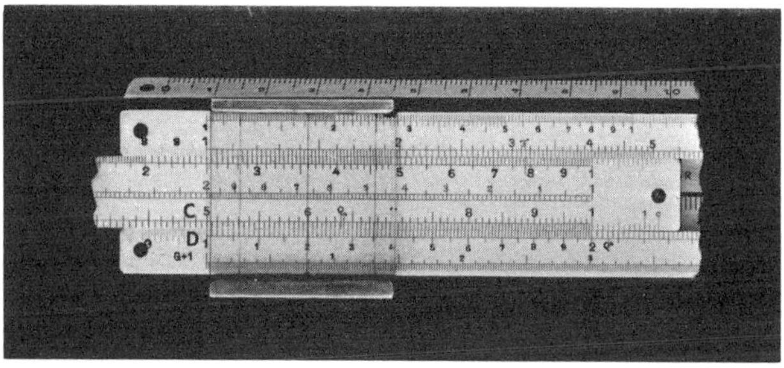

Figure A5

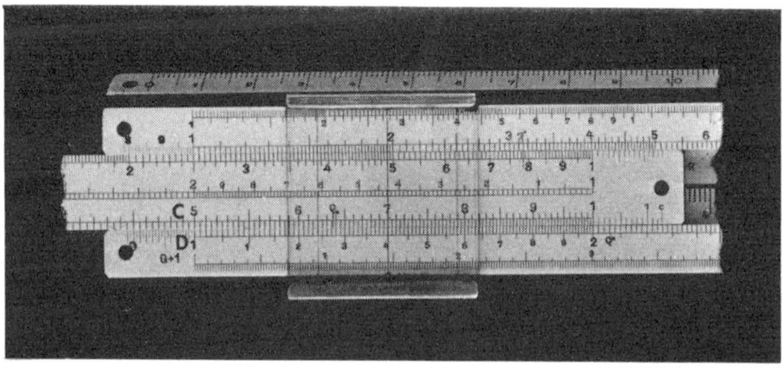

Figure A6

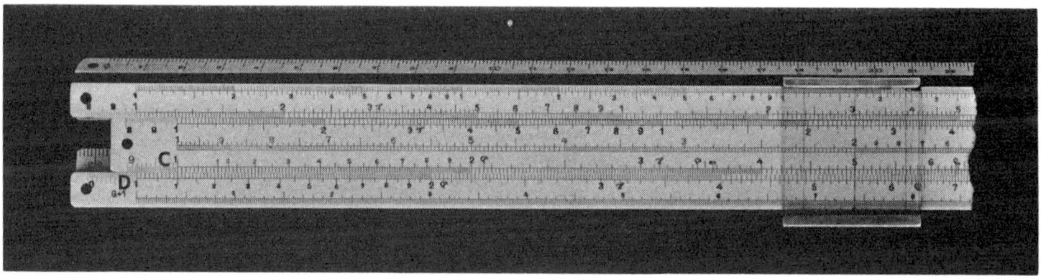

Figure A7

It is the largest mark between 5 and 6 and it is the *fifth* large mark after the 5. (There are some shorter marks between the large marks, which you may ignore for the time being.) This mark is equivalent to a small 5, but for lack of space small 5 is not labeled as such. Hence, you read under the hairline on the body scale D "big 5 small 5," i.e., 55.

3. $54 \div 9 = 6$. Set hairline on "big 5 small 4" of body, i.e., on the fourth large mark after the big 5 (Fig. A8). Set the 9 of the slide under the hairline. Read the quotient 6 on the body under the right 1 of the slide.

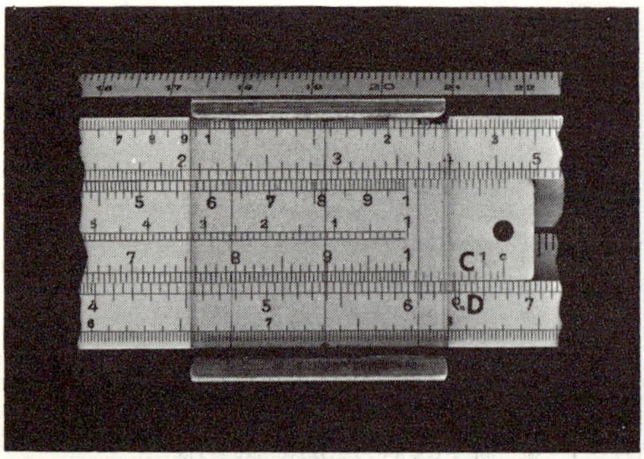

Figure A8

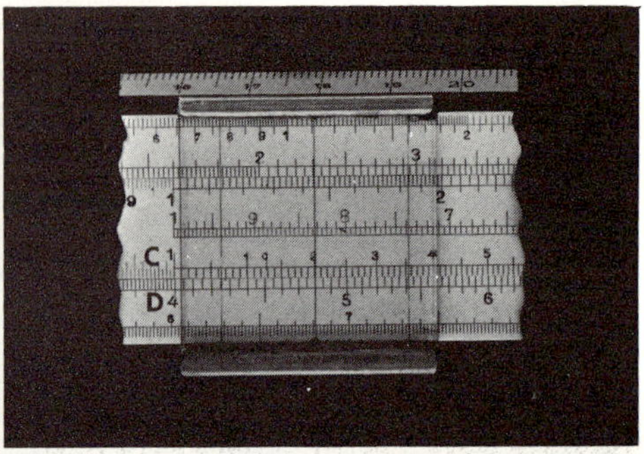

Figure A9

4. 48 ÷ 12 = 4. Set hairline on the 8th large mark between 4 and 5, i.e., on "big 4 small 8" of the body (Fig. A9). Set "large 1 small 2" of slide under hairline. Read quotient 4 on body under left 1 of slide.

You now know how to multiply and divide on a slide rule any two-digit numbers. But you may want to notice that as many as 3 or even 4 digits may be read on your slide rule. For example, if you move the hairline past big 4 (Fig. A10) on the body to the *smaller* mark between "big 4 small 5" and "big 4 small 6," i.e., between the fifth and sixth large marks after the 4, you read the number 4, 5, 5 (get into the habit of reading "four, five, five," since you do not know where the decimal point is). Similarly, you read 8, 7, 5 (eight, seven, five) when the hairline is on the smaller mark between the seventh and the eighth large marks after the big 8 on the body or on the slide (Fig. A11). Thus, you obtain 455 ÷ 875 = 0.52 (Fig. A11) by setting 8, 7, 5

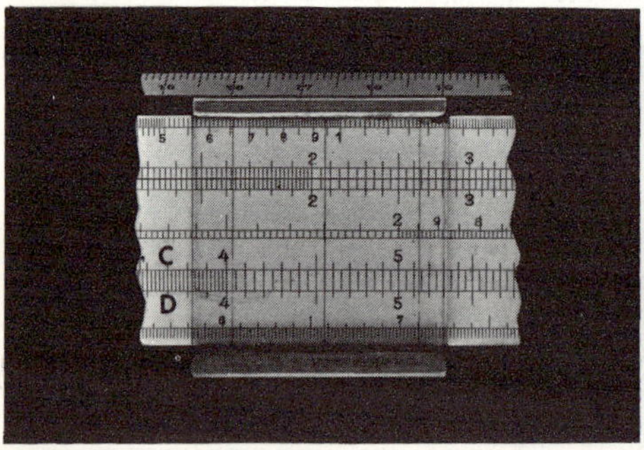

Figure A10

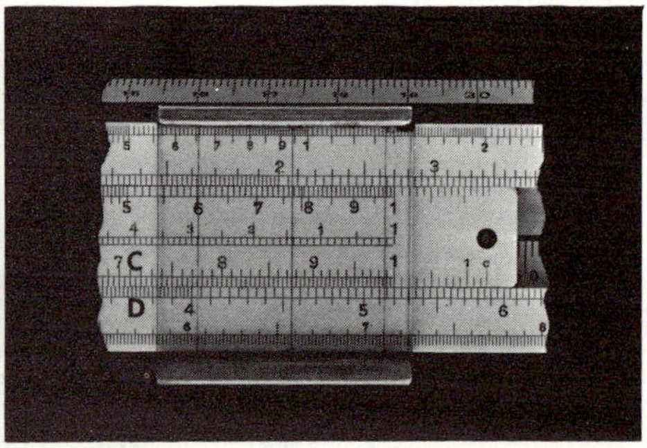

Figure A11

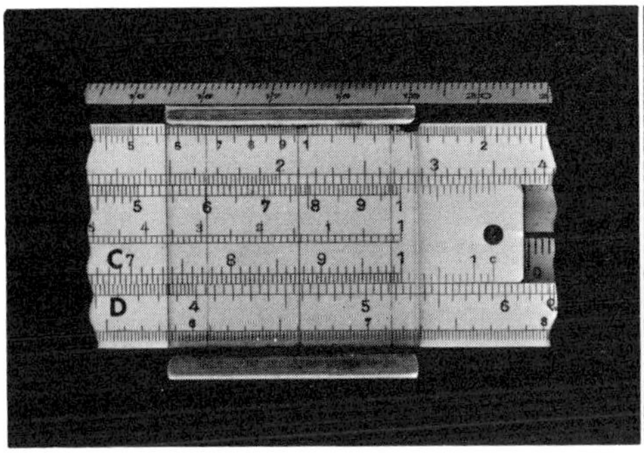

Figure A12a

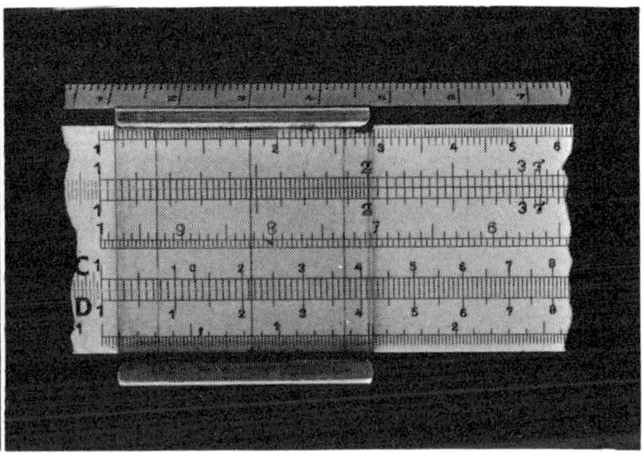

Figure A12b

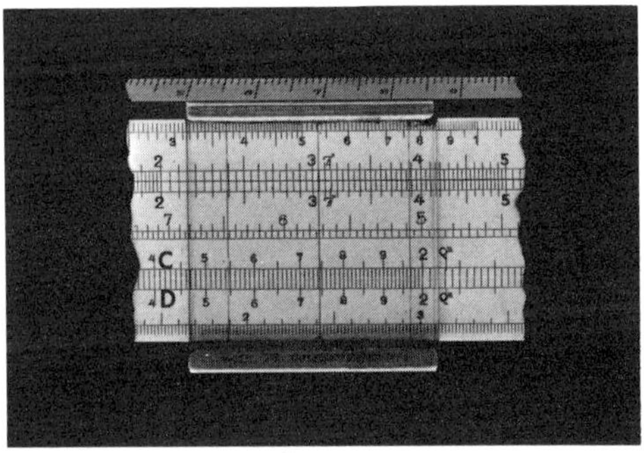

Figure A12c

of the slide over 4, 5, 5 of the body and by reading the quotient 5, 2 on the rule under the right 1 of the slide.

Finally, you may "estimate" a number like 4, 5, 8 on the body or the slide scales by setting the hairline between 4, 5, 5 and 4, 6, nearer to 4, 6 than to 4, 5, 5 (Fig. A12a). Dividing 4, 5, 8 by 8, 7, 5 you may also "estimate" the quotient as 5, 2, 3, since it is less than 5, 2, 5, more than 5, 2, 0 and nearer to 5, 2, 5 than to 5, 2, 0: $458 \div 875 = 0.523$.

Between big 1 and big 2 on either scale D or C, you can read exactly the numbered marks 11, 12, . . . , 19; you can also read exactly on the smaller unnumbered marks numbers like 1, 2, 1 or 1, 7, 4, and you can even estimate numbers like 1, 2, 1, 5 (Fig. A12b) or 1, 7, 4, 5 (Fig. A12c) by setting the hairline halfway between the marks 1, 2, 1 and 1, 2, 2, or between 1, 7, 4 and 1, 7, 5.

There are two more scales you must learn to use to complete your elementary slide rule requirements: the square root or square scale (scale

A) and the cube root or cube scale (scale K) (Fig. A1). The A scale appears on the upper part of the body; the K scale appears either above A or below D. The A and K scales *do not require the use of the slide*, but only of the hairline.

The square root scale A is the lower scale on the upper part of the body whenever the K scale is in this part of the body. You will recognize it because it is divided into *2 equal parts*, marked 1, 2, 3, . . . , 9, 1, 2, 3, . . . , 9, 1 (Fig. A1). By setting the hairline on a number on this scale one reads directly its square root on the body scale D (the scale used in multiplication and division). Numbers from *1 to 10* are read on the *left* half of the scale; numbers from *10 to 100* are read on the *right* half. For example, set the hairline on the left 4 of the square root scale A and you read below on the body scale D its square root 2 (Fig. A13). Set the hairline on the right 4 of the square root scale A, that is on 40, and you read (or estimate) 6.32 (Fig. A14) below on the body scale D. By the "powers of 10" method, you then find:

$$\sqrt{400} = \sqrt{4 \times 100} = \sqrt{4 \times 10^2} = \sqrt{4} \times \sqrt{100}$$
$$= 10\sqrt{4} = 10 \times 2 = 20,$$
$$\sqrt{4{,}000} = \sqrt{40 \times 10^2} = 10\sqrt{40} = 10 \times 6.32 = 63.2,$$
$$\sqrt{40{,}000{,}000} = \sqrt{4 \times 10^6} = 10^3\sqrt{4} = 10^3 \times 2 = 2{,}000.$$

Notice now that if you read a number on the body scale D, its square appears directly above on the square root scale A. (This is why scale A is also called the square scale.) For example, read 4 on the body scale D and 16 above it on the right half of the square scale A (Fig. A15).

The cube root scale K is above the square root scale A in most slide rules and is recognizable because it is divided into *3 equal parts*. Numbers from *1 to 10* are read on the *left* part, from *10 to 100* on the *middle* part, from

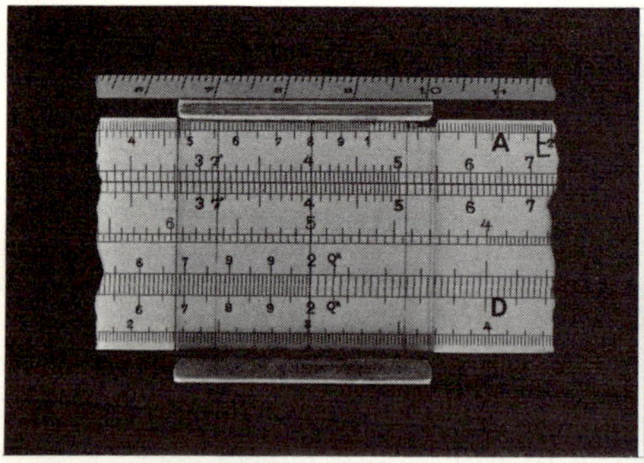

Figure A13

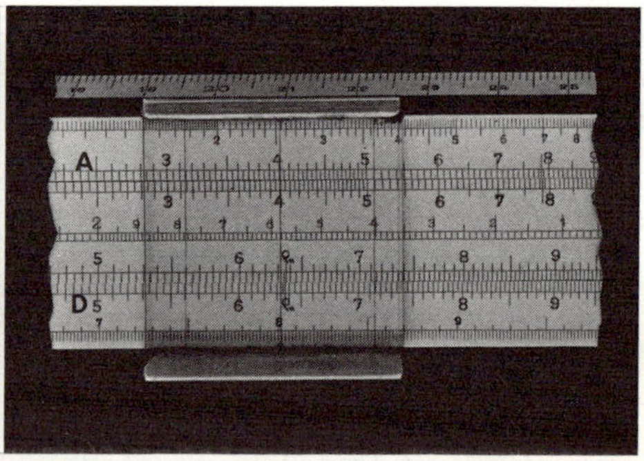

Figure A14

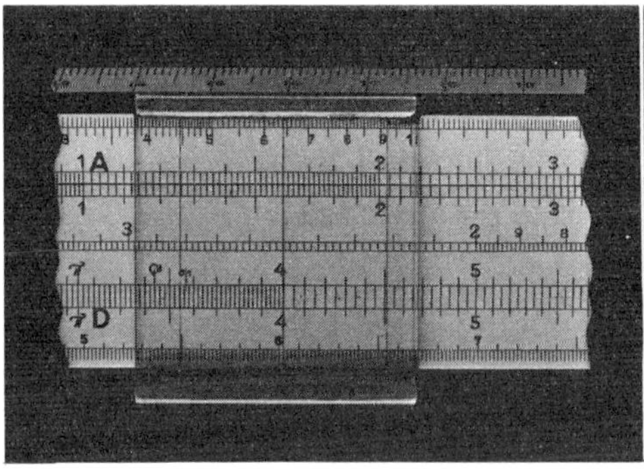

Figure A15

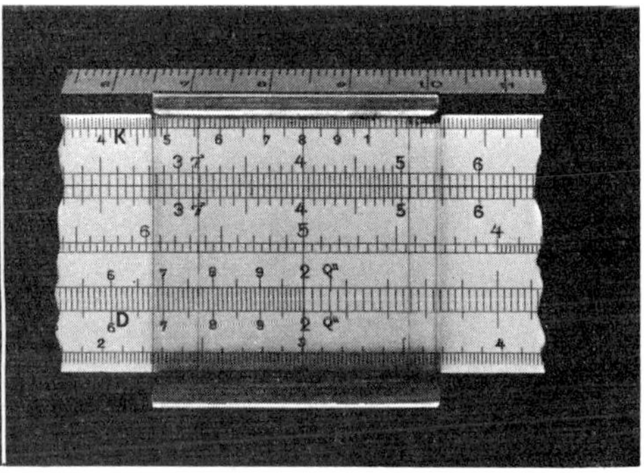

Figure A16

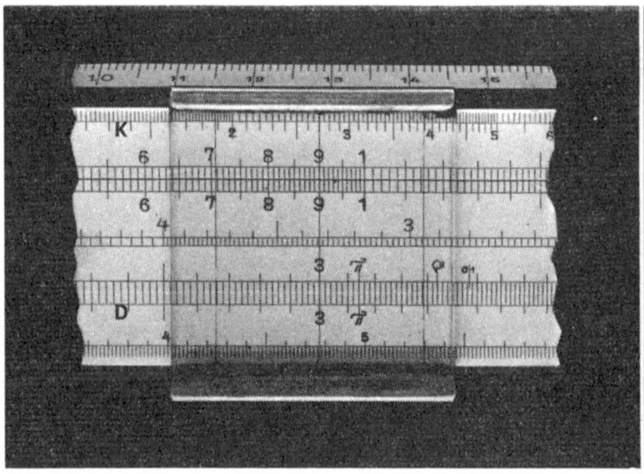

Figure A17

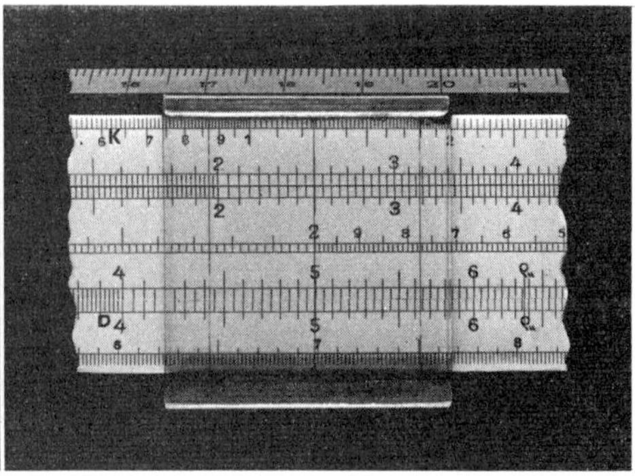

Figure A18

100 to 1,000 on the *right* part of scale K; their cube roots appear on the body scale D (Fig. A1). For example, read 8 on the left part of K and $2 = \sqrt[3]{8}$ below on the body scale D (Fig. A16). Read 27 on the middle part of K and $3 = \sqrt[3]{27}$ below on D (Fig. A17). Read 125 on the right part of K and $5 = \sqrt[3]{125}$ below on D (Fig. A18). By the "powers of 10" rule you then find:

$$\sqrt[3]{0.8} = \sqrt[3]{\frac{800}{1{,}000}} = \sqrt[3]{\frac{800}{10^3}} = \frac{\sqrt[3]{800}}{\sqrt[3]{10^3}}$$

$$= \frac{1}{10}\sqrt[3]{800} = \frac{1}{10}9.3 = 0.93 \quad \text{(using right part of K).}$$

$$\sqrt[3]{2{,}700} = \sqrt[3]{2.7 \times 10^3} = 10\sqrt[3]{2.7}$$

$$= 10 \times 1.39 = 13.9 \quad \text{(using left part of K).}$$

$$\sqrt[3]{0.08} = \sqrt[3]{80 \times 10^{-3}} = 10^{-1}\sqrt[3]{80}$$

$$= 10^{-1} \times 4.3 = 0.43 \quad \text{(using middle part of K).}$$

Inversely, read 5 on the body scale D and $125 = 5^3$ above on K (Fig. A18); 3 on the body scale D and $27 = 3^3$ above on K (Fig. A17); 2 on the body scale D and $8 = 2^3$ above on K (Fig. A16).

Exercise on the 4 scales you have studied so far and you will know *all* that is strictly necessary in order to use a slide rule. Later on you may want to use all the other scales (there are many on both faces of the slide rule) and become an expert by reading the instruction booklet that comes with your slide rule. But *first familiarize yourself with the use of the 4 basic scales* (C, D, A, and K) and, above all, with the "powers of 10" rule.

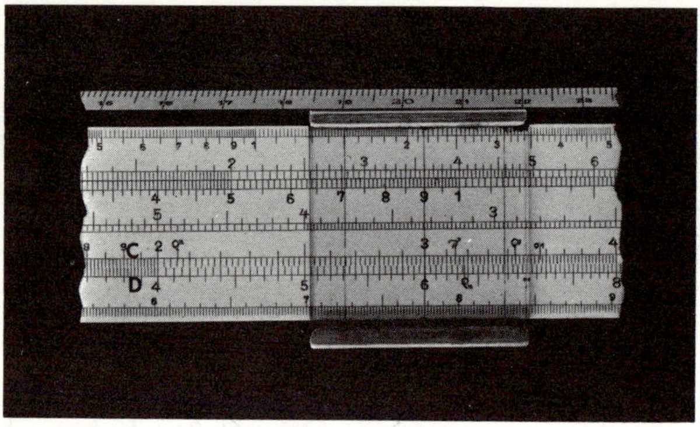

Figure A19

Actually you may want to learn one more slide rule procedure by simply noticing that chain operations often require only one setting of the slide. This will save you unnecessary writing of intermediate numbers on paper. For example, the operation $\frac{4}{2} \times 3$ requires the setting of the 2 of the slide over the 4 of the body, obtaining a quotient 2 under the left 1 of the slide. To multiply $\frac{4}{2} = 2$ by 3, the left 1 of the slide must be set on the 2 of the body, but since *it is already there*, all you have to do is set the hairline on 3 of the slide and read 6 on the body (Fig. A19). Perform with only two settings of the slide the following operations: $(4 \times 3)/(2 \times 5)$, noticing that divisions must be performed first. In other words, perform the operations in the following order:

$$\frac{4 \times 3}{2 \times 5} = \left(\frac{4}{2}\right) \times \frac{3}{5} = 2 \times \frac{3}{5} = \frac{6}{5} = 1.2.$$

1.2 is read on the body scale D under the left 1 of the slide (Fig. A20a, b, c) and the intermediate numbers $2 = \frac{4}{2}$ and $6 = 2 \times 3$ need not be written or *even noticed.*

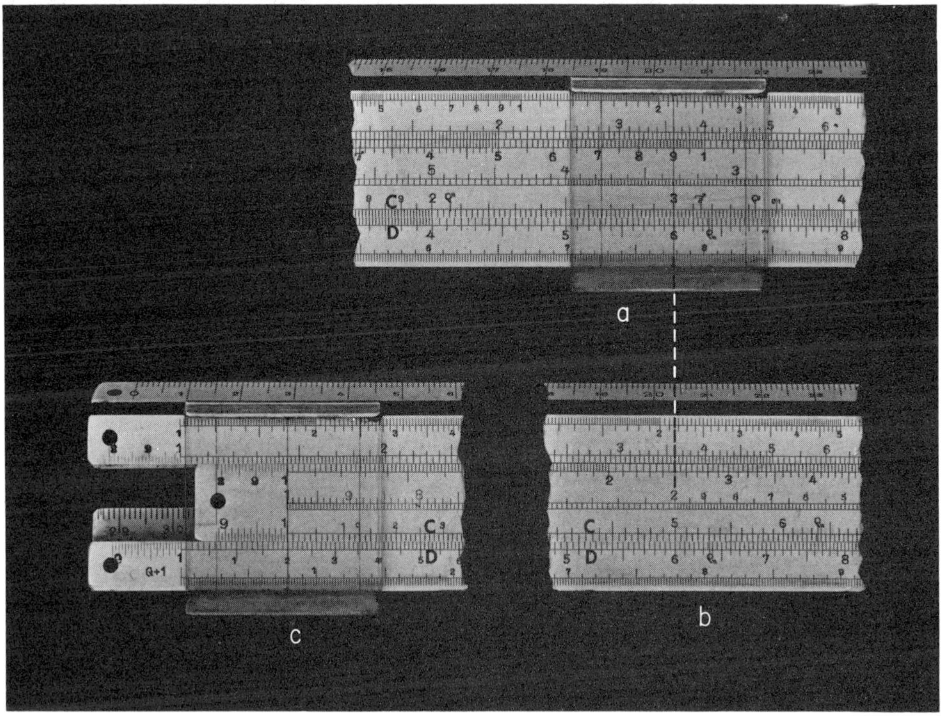

Figure A20

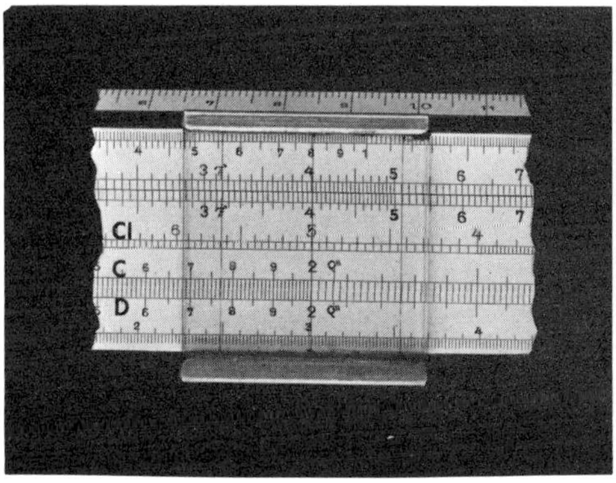

Figure A21a

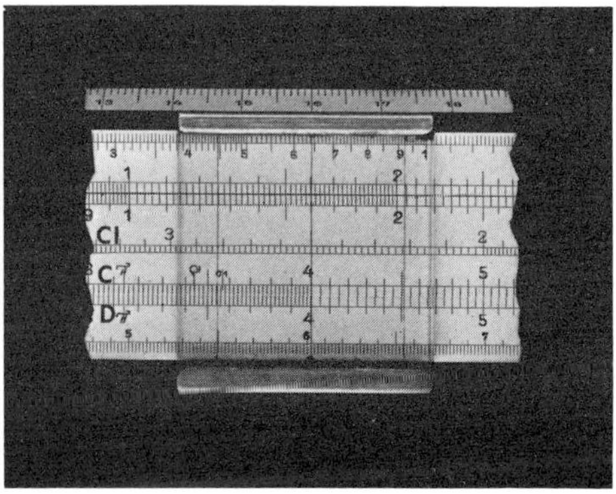

Figure A21b

Finally, you will notice on the slide a scale marked *from right to left* 1, 2, 3, . . . , 9, 1 (and sometimes marked with *red* numbers). This is the *reciprocal scale* CI (Fig. A1). By setting the hairline on a number N of the slide scale C, you can read directly its reciprocal $1/N$ on scale CI. For example, you read 5 on CI above 2 on C, since $\frac{1}{2} = 0.5$ (Fig. A21a). You read 2.5 on CI over 4 on C, since $\frac{1}{4} = 0.25$ (Fig. A21b).

The reciprocal scale is particularly useful in multiplication and division, since multiplication by $1/N$ is equivalent to division by N and division by $1/N$ is equivalent to multiplication by N. The reciprocal scale CI is often used to avoid shifting the slide when the left or right 1 falls outside the body.

You should have spent 2 to 3 hours in studying this section of the Appendix. From now on you need only to practice and to acquire *confidence* in using the slide rule. Once you have gained confidence you will not waste time checking slide rule operations longhand, as you are almost bound to do at first.

A.2 FORMULAS

Chapter 1. Arithmetic

$\ldots,\ 10^{-4} = 0.0001,\ 10^{-3} = 0.001,\ 10^{-2} = 0.01,\ 10^{-1} = 0.1,\ 10^0 = 1,\ 10^1 = 10,\ 10^2 = 100,\ 10^3 = 1{,}000,\ 10^4 = 10{,}000, \ldots.$

$\ldots,\ 2{,}740 = 2.74 \times 10^3,\ 274 = 2.74 \times 10^2,\ 27.4 = 2.74 \times 10^1,\ 2.74 = 2.74 \times 10^0,\ 0.274 = 2.74 \times 10^{-1},\ 0.0274 = 2.74 \times 10^{-2},\ 0.00274 = 2.74 \times 10^{-3}, \ldots.$

Chapter 2. Algebra

$$a^2 - b^2 = (a + b)(a - b) \qquad (2.2.3)$$

$$a^m \times a^n = a^{m+n} \qquad (2.3.1)$$

$$(a^m)^n = (a^n)^m = a^{m \times n} \qquad (2.3.2)$$

$$\sqrt[n]{a} = a^{1/n} \qquad (2.3.3)$$

$$\frac{1}{a^n} = a^{-n} \qquad (2.3.4)$$

$$\frac{1}{\sqrt[n]{a}} = a^{-1/n} \tag{2.3.5}$$

$$(a \pm b)^2 = a^2 \pm 2ab + b^2 \tag{2.4.1}$$

$$(a \pm b)^3 = a^3 \pm 3a^2b + 3ab^2 \pm b^3 \tag{2.4.2}$$

$$(a \pm b)^4 = a^4 \pm 4a^3b + 6a^2b^2 \pm 4ab^3 + b^4 \tag{2.4.3}$$

$$(a \pm b)^n = a^n \pm \frac{n}{1}a^{n-1}b + \frac{n(n-1)}{1\cdot 2}a^{n-2}b^2 \pm \frac{n(n-1)(n-2)}{1\cdot 2\cdot 3}a^{n-3}b^3 + \ldots + (-1)^{n-1}\frac{n}{1}ab^{n-1} + (-1)^n b^n \tag{2.4.4}$$

$$(a \pm b)^n \simeq a^n\left(1 \pm n\frac{b}{a}\right) \qquad (b \ll a) \tag{2.4.7}$$

$$\frac{1}{a \pm b} \simeq \frac{1}{a}\left(1 \mp \frac{b}{a}\right) \qquad (b \ll a) \tag{2.4.8}$$

$$\sqrt{a \pm b} \simeq \sqrt{a}\left(1 \pm \frac{1}{2}\frac{b}{a}\right) \qquad (b \ll a) \tag{2.4.9}$$

$$\sqrt[3]{a \pm b} = \sqrt[3]{a}\left(1 \pm \frac{1}{3}\frac{b}{a}\right) \qquad (b \ll a) \tag{2.4.10}$$

$$ax^2 + bx + c = 0; \qquad x_{1,2} = \frac{1}{2a}(-b \pm \sqrt{b^2 - 4ac}) \tag{2.6.2}$$

CHAPTER 3. TRIGONOMETRY

$$\sin\alpha = \frac{S}{R}, \quad \cos\alpha = \frac{C}{R}, \quad \tan\alpha = \frac{T}{R} = \frac{S}{C}$$
$$\cot\alpha = \frac{R}{T} = \frac{C}{S} = \frac{1}{\tan\alpha} \tag{3.2.1}$$

$$\sin^2\alpha + \cos^2\alpha = 1 \tag{3.4.1}$$

$$\sin\alpha = \sqrt{1 - \cos^2\alpha}, \qquad \tan\alpha = \frac{\sqrt{1 - \cos^2\alpha}}{\cos\alpha} \tag{3.4.2}$$

$$\cos\alpha = \sqrt{1 - \sin^2\alpha}, \qquad \tan\alpha = \frac{\sin\alpha}{\sqrt{1 - \sin^2\alpha}} \tag{3.4.3}$$

$$\sin\alpha = \frac{\tan\alpha}{\sqrt{1 - \tan^2\alpha}}, \qquad \cos\alpha = \frac{1}{\sqrt{1 - \tan^2\alpha}} \tag{3.4.4, 5}$$

$$\sin(90° - \alpha) = \cos\alpha, \quad \cos(90° - \alpha) = \sin\alpha, \quad \tan(90° - \alpha) = \frac{1}{\tan\alpha} \tag{3.4.6}$$

$$\sin(\alpha \pm \beta) = \sin\alpha\cos\beta \pm \cos\alpha\sin\beta \tag{3.4.8}$$

$$\cos(\alpha \pm \beta) = \cos\alpha\cos\beta \mp \sin\alpha\sin\beta \tag{3.4.9}$$

$$\tan(\alpha \pm \beta) = \frac{\tan\alpha \pm \tan\beta}{1 \mp \tan\alpha\tan\beta} \tag{3.4.10}$$

$$\sin 2\alpha = 2\sin\alpha\cos\alpha, \quad \cos 2\alpha = \cos^2\alpha - \sin^2\alpha, \quad \tan 2\alpha = \frac{2\tan\alpha}{1 - \tan^2\alpha} \tag{3.4.11}$$

$$\sin(180° \mp \alpha) = \pm\sin\alpha, \quad \cos(180° \mp \alpha) = -\cos\alpha, \quad \tan(180° \mp \alpha) = \mp\tan\alpha \tag{3.5.2}$$

$$\begin{aligned} \sin(360° - \alpha) &= \sin(-\alpha) = -\sin\alpha, \\ \cos(360° - \alpha) &= \cos(-\alpha) = \cos\alpha, \\ \tan(360° - \alpha) &= \tan(-\alpha) = -\tan\alpha \end{aligned} \tag{3.5.4}$$

$$\alpha° = 57.296\, x \text{ rad} \tag{3.6.1}$$

$$\sin x = x, \quad \cos x = 1, \quad \tan x = x \qquad (x < 1) \tag{3.7.1}$$

$$\sin x = x - \frac{x^3}{6}, \quad \cos x = 1 - \frac{x^2}{2}, \quad \tan x = x + \frac{x^3}{3} \qquad (x \ll 1) \tag{3.7.2, 3, 4}$$

Chapter 4. Analytic Geometry

$$P(x, y): \quad OP = \sqrt{x^2 + y^2}, \qquad \tan\theta = \frac{y}{x} \tag{4.2.1, 2}$$

$$P_1(x_1, y_1), P_2(x_2, y_2): \quad d = \sqrt{(x_1 - x_2)^2 + (y_1 - y_2)^2} \tag{4.3.1}$$

$$\tan\theta = \frac{y_2 - y_1}{x_2 - x_1} \tag{4.3.2}$$

$$\text{Line through } P_1P_2: \quad y = y_1 + \frac{y_2 - y_1}{x_2 - x_1}(x - x_1) \tag{4.4.1}$$

$$\text{Line through } P_1 \text{ with slope } m: y = y_1 + m(x - x_1) \tag{4.4.2}$$

General equation of straight line:

$$ay + bx + c = 0, \quad m = -\frac{b}{a}, \quad y_i = -\frac{c}{a}, \quad x_i = -\frac{c}{b} \qquad (4.4.6, 8, 9)$$

Linear interpolation formulas:

$$y_3 = y_1 + \frac{x_3 - x_1}{x_2 - x_1}(y_2 - y_1) \qquad (4.5.1)$$

$$x_3 = x_1 + \frac{y_3 - y_1}{y_2 - y_1}(x_2 - x_1) \qquad (4.5.2)$$

$$\bar{x}_3 = \frac{y_2 x_1 - y_1 x_2}{y_2 - y_1} \qquad (4.5.3)$$

Chapter 5. Differential Calculus

Function	*Derivative*	
x^n	nx^{n-1}	(5.3.4)
constant	0	(5.3.6)
$\sin x$	$\cos x$	(5.4.1)
$\cos x$	$-\sin x$	(5.4.2)
$\tan x$	$\frac{1}{\cos^2 x}$	(5.4.3)
$\ln x$	$\frac{1}{x}$	(6.6.4)
$e^{\pm x}$	$\pm e^{\pm x}$	(6.8.3)
cy	$c\frac{dy}{dx}$	(5.3.5)
$u \times v$	$u\frac{dv}{dx} + v\frac{du}{dx}$	(5.7.3)
$\frac{u}{v}$	$\frac{v\,(du/dx) - u\,(dv/dx)}{v^2}$	(5.7.5)
$y = f_1(z),\ z = f_2(x)$	$\frac{dy}{dx} = \frac{dy}{dz}\frac{dz}{dx}$	(5.7.7)

CHAPTER 6. INTEGRAL CALCULUS

Function	*Integral*	
x^n	$\frac{1}{n+1}x^{n+1} + C \qquad (n \neq 1)$	(6.6.1)
$\frac{1}{x}$	$\ln x + C$	(6.6.4)
$\sin x$	$-\cos x + C$	(6.6.3)
$\cos x$	$\sin x + C$	(6.6.2)
$\tan x$	$-\ln(\cos x) + C$	(6.6.5)
$e^{\pm x}$	$\pm e^{\pm x} + C$	(6.7.4)

Table 1. SQUARE AND CUBE ROOTS

x	$\sqrt{x}$	$\sqrt{10x}$	$\sqrt[3]{x}$	$\sqrt[3]{10x}$	$\sqrt[3]{100x}$
1	1.000 000	3.162 278	1.000 000	2.154 435	4.641 589
2	1.414 214	4.472 136	1.259 921	2.714 418	5.848 035
3	1.732 051	5.477 226	1.442 250	3.107 233	6.694 330
4	2.000 000	6.324 555	1.587 401	3.419 952	7.368 063
5	2.236 068	7.071 068	1.709 976	3.684 031	7.937 005
6	2.449 490	7.745 967	1.817 121	3.914 868	8.434 327
7	2.645 751	8.366 600	1.912 931	4.121 285	8.879 040
8	2.828 427	8.944 272	2.000 000	4.308 869	9.283 178
9	3.000 000	9.486 833	2.080 084	4.481 405	9.654 894
10	3.162 278	10.00000	2.154 435	4.641 589	10.00000
11	3.316 625	10.48809	2.223 980	4.791 420	10.32280
12	3.464 102	10.95445	2.289 428	4.932 424	10.62659
13	3.605 551	11.40175	2.351 335	5.065 797	10.91393
14	3.741 657	11.83216	2.410 142	5.192 494	11.18689
15	3.872 983	12.24745	2.466 212	5.313 293	11.44714
16	4.000 000	12.64911	2.519 842	5.428 835	11.69607
17	4.123 106	13.03840	2.571 282	5.539 658	11.93483
18	4.242 641	13.41641	2.620 741	5.646 216	12.16440
19	4.358 899	13.78405	2.668 402	5.748 897	12.38562
20	4.472 136	14.14214	2.714 418	5.848 035	12.59921
21	4.582 576	14.49138	2.758 924	5.943 922	12.80579
22	4.690 416	14.83240	2.802 039	6.036 811	13.00591
23	4.795 832	15.16575	2.843 867	6.126 926	13.20006
24	4.898 979	15.49193	2.884 499	6.214 465	13.38866
25	5.000 000	15.81139	2.924 018	6.299 605	13.57209
26	5.099 020	16.12452	2.962 496	6.382 504	13.75069
27	5.196 152	16.43168	3.000 000	6.463 304	13.92477
28	5.291 503	16.73320	3.036 589	6.542 133	14.09460
29	5.385 165	17.02939	3.072 317	6.619 106	14.26043
30	5.477 226	17.32051	3.107 233	6.694 330	14.42250
31	5.567 764	17.60682	3.141 381	6.767 899	14.58100
32	5.656 854	17.88854	3.174 802	6.839 904	14.73613
33	5.744 563	18.16590	3.207 534	6.910 423	14.88806
34	5.830 952	18.43909	3.239 612	6.979 532	15.03695
35	5.916 080	18.70829	3.271 066	7.047 299	15.18294
36	6.000 000	18.97367	3.301 927	7.113 787	15.32619
37	6.082 763	19.23538	3.332 222	7.179 054	15.46680
38	6.164 414	19.49359	3.361 975	7.243 156	15.60491
39	6.244 998	19.74842	3.391 211	7.306 144	15.74061
40	6.324 555	20.00000	3.419 952	7.368 063	15.87401
41	6.403 124	20.24846	3.448 217	7.428 959	16.00521
42	6.480 741	20.49390	3.476 027	7.488 872	16.13429
43	6.557 439	20.73644	3.503 398	7.547 842	16.26133
44	6.633 250	20.97618	3.530 348	7.605 905	16.38643
45	6.708 204	21.21320	3.556 893	7.663 094	16.50964
46	6.782 330	21.44761	3.583 048	7.719 443	16.63103
47	6.855 655	21.67948	3.608 826	7.774 980	16.75069
48	6.928 203	21.90890	3.634 241	7.829 735	16.86865
49	7.000 000	22.13594	3.659 306	7.883 735	16.98499
50	7.071 068	22.36068	3.684 031	7.937 005	17.09976

Table 1. (Cont.)

x	$\sqrt{x}$	$\sqrt{10x}$	$\sqrt[3]{x}$	$\sqrt[3]{10x}$	$\sqrt[3]{100x}$
50	7.071 068	22.36068	3.684 031	7.937 005	17.09976
51	7.141 428	22.58318	3.708 430	7.989 570	17.21301
52	7.211 103	22.80351	3.732 511	8.041 452	17.32478
53	7.280 110	23.02173	3.756 286	8.092 672	17.43513
54	7.348 469	23.23790	3.779 763	8.143 253	17.54411
55	7.416 198	23.45208	3.802 952	8.193 213	17.65174
56	7.483 315	23.66432	3.825 862	8.242 571	17.75808
57	7.549 834	23.87467	3.848 501	8.291 344	17.86316
58	7.615 773	24.08319	3.870 877	8.339 551	17.96702
59	7.681 146	24.28992	3.892 996	8.387 207	18.06969
60	7.745 967	24.49490	3.914 868	8.434 327	18.17121
61	7.810 250	24.69818	3.936 497	8.480 926	18.27160
62	7.874 008	24.89980	3.957 892	8.527 019	18.37091
63	7.937 254	25.09980	3.979 057	8.572 619	18.46915
64	8.000 000	25.29822	4.000 000	8.617 739	18.56636
65	8.062 258	25.49510	4.020 726	8.662 391	18.66256
66	8.124 038	25.69047	4.041 240	8.706 588	18.75777
67	8.185 353	25.88436	4.061 548	8.750 340	18.85204
68	8.246 211	26.07681	4.081 655	8.793 659	18.94536
69	8.306 624	26.26785	4.101 566	8.836 556	19.03778
70	8.366 600	26.45751	4.121 285	8.879 040	19.12931
71	8.426 150	26.64583	4.140 818	8.921 121	19.21997
72	8.485 281	26.83282	4.160 168	8.962 809	19.30979
73	8.544 004	27.01851	4.179 339	9.004 113	19.39877
74	8.602 325	27.20294	4.198 336	9.045 042	19.48695
75	8.660 254	27.38613	4.217 163	9.085 603	19.57434
76	8.717 798	27.56810	4.235 824	9.125 805	19.66095
77	8.774 964	27.74887	4.254 321	9.165 656	19.74681
78	8.831 761	27.92848	4.272 659	9.205 164	19.83192
79	8.888 194	28.10694	4.290 840	9.244 335	19.91632
80	8.944 272	28.28427	4.308 869	9.283 178	20.00000
81	9.000 000	28.46050	4.326 749	9.321 698	20.08299
82	9.055 385	28.63564	4.344 481	9.359 902	20.16530
83	9.110 434	28.80972	4.362 071	9.397 796	20.24694
84	9.165 151	28.98275	4.379 519	9.435 388	20.32793
85	9.219 544	29.15476	4.396 830	9.472 682	20.40828
86	9.273 618	29.32576	4.414 005	9.509 685	20.48800
87	9.327 379	29.49576	4.431 048	9.546 403	20.56710
88	9.380 832	29.66479	4.447 960	9.582 840	20.64560
89	9.433 981	29.83287	4.464 745	9.619 002	20.72351
90	9.486 833	30.00000	4.481 405	9.654 894	20.80084
91	9.539 392	30.16621	4.497 941	9.690 521	20.87759
92	9.591 663	30.33150	4.514 357	9.725 888	20.95379
93	9.643 651	30.49590	4.530 655	9.761 000	21.02944
94	9.695 360	30.65942	4.546 836	9.795 861	21.10454
95	9.746 794	30.82207	4.562 903	9.830 476	21.17912
96	9.797 959	30.98387	4.578 857	9.864 848	21.25317
97	9.848 858	31.14482	4.594 701	9.898 983	21.32671
98	9.899 495	31.30495	4.610 436	9.932 884	21.39975
99	9.949 874	31.46427	4.626 065	9.966 555	21.47229
100	10.00000	31.62278	4.641 589	10.00000	21.54435

Table 2. TRIGONOMETRIC FUNCTIONS

Radians	Degrees	Sine	Cosine	Tangent	Cotangent		
0.0000	0°00′	.0000	1.0000	.0000	∞	90°00′	1.5708
0.0029	10	.0029	1.0000	.0029	343.77	50	1.5679
0.0058	20	.0058	1.0000	.0058	171.89	40	1.5650
0.0087	30	.0087	1.0000	.0087	114.59	30	1.5621
0.0116	40	.0116	.9999	.0116	85.940	20	1.5592
0.0145	50	.0145	.9999	.0145	68.750	10	1.5563
0.0175	1°00′	.0175	.9998	.0175	57.290	89°00′	1.5533
0.0204	10	.0204	.9998	.0204	49.104	50	1.5504
0.0233	20	.0233	.9997	.0233	42.964	40	1.5475
0.0262	30	.0262	.9997	.0262	38.188	30	1.5446
0.0291	40	.0291	.9996	.0291	34.368	20	1.5417
0.0320	50	.0320	.9995	.0320	31.242	10	1.5388
0.0349	2°00′	.0349	.9994	.0349	28.636	88°00′	1.5359
0.0378	10	.0378	.9993	.0378	26.432	50	1.5330
0.0407	20	.0407	.9992	.0407	24.542	40	1.5301
0.0436	30	.0436	.9990	.0437	22.904	30	1.5272
0.0465	40	.0465	.9989	.0466	21.470	20	1.5243
0.0495	50	.0494	.9988	.0495	20.206	10	1.5213
0.0524	3°00′	.0523	.9986	.0524	19.081	87°00′	1.5184
0.0553	10	.0552	.9985	.0553	18.075	50	1.5155
0.0582	20	.0581	.9983	.0582	17.169	40	1.5126
0.0611	30	.0610	.9981	.0612	16.350	30	1.5097
0.0640	40	.0640	.9980	.0641	15.605	20	1.5068
0.0669	50	.0669	.9978	.0670	14.924	10	1.5039
0.0698	4°00′	.0698	.9976	.0699	14.301	86°00′	1.5010
0.0727	10	.0727	.9974	.0729	13.727	50	1.4981
0.0756	20	.0756	.9971	.0758	13.197	40	1.4952
0.0785	30	.0785	.9969	.0787	12.706	30	1.4923
0.0814	40	.0814	.9967	.0816	12.251	20	1.4893
0.0844	50	.0843	.9964	.0846	11.826	10	1.4864
0.0873	5°00′	.0872	.9962	.0875	11.430	85°00′	1.4835
0.0902	10	.0901	.9959	.0904	11.059	50	1.4806
0.0931	20	.0929	.9957	.0934	10.712	40	1.4777
0.0960	30	.0958	.9954	.0963	10.385	30	1.4748
0.0989	40	.0987	.9951	.0992	10.078	20	1.4719
0.1018	50	.1016	.9948	.1022	9.7882	10	1.4690
0.1047	6°00′	.1045	.9945	.1051	9.5144	84°00′	1.4661
0.1076	10	.1074	.9942	.1080	9.2553	50	1.4632
0.1105	20	.1103	.9939	.1110	9.0098	40	1.4603
0.1134	30	.1132	.9936	.1139	8.7769	30	1.4574
0.1164	40	.1161	.9932	.1169	8.5555	20	1.4544
0.1193	50	.1190	.9929	.1198	8.3450	10	1.4515
0.1222	7°00′	.1219	.9925	.1228	8.1443	83°00′	1.4486
0.1251	10	.1248	.9922	.1257	7.9530	50	1.4457
0.1280	20	.1276	.9918	.1287	7.7704	40	1.4428
0.1309	30	.1305	.9914	.1317	7.5958	30	1.4399
0.1338	40	.1334	.9911	.1346	7.4287	20	1.4370
0.1367	50	.1363	.9907	.1376	7.2687	10	1.4341
0.1396	8°00′	.1392	.9903	.1405	7.1154	82°00′	1.4312
0.1425	10	.1421	.9899	.1435	6.9682	50	1.4283
0.1454	20	.1449	.9894	.1465	6.8269	40	1.4254
0.1484	30	.1478	.9890	.1495	6.6912	30	1.4224
0.1513	40	.1507	.9886	.1524	6.5606	20	1.4195
0.1542	50	.1536	.9881	.1554	6.4348	10	1.4166
0.1571	9°00′	.1564	.9877	.1584	6.3138	81°00′	1.4137
		Cosine	Sine	Cotangent	Tangent	Degrees	Radians

Table 2. (CONT.)

Radians	Degrees	Sine	Cosine	Tangent	Cotangent		
0.1571	9°00′	.1564	.9877	.1584	6.3138	81°00′	1.4137
0.1600	10	.1593	.9872	.1614	6.1970	50	1.4108
0.1629	20	.1622	.9868	.1644	6.0844	40	1.4079
0.1658	30	.1650	.9863	.1673	5.9758	30	1.4050
0.1687	40	.1679	.9858	.1703	5.8708	20	1.4021
0.1716	50	.1708	.9853	.1733	5.7694	10	1.3992
0.1745	10°00′	.1736	.9848	.1763	5.6713	80°00′	1.3963
0.1774	10	.1765	.9843	.1793	5.5764	50	1.3934
0.1804	20	.1794	.9838	.1823	5.4845	40	1.3904
0.1833	30	.1822	.9833	.1853	5.3955	30	1.3875
0.1862	40	.1851	.9827	.1883	5.3093	20	1.3846
0.1891	50	.1880	.9822	.1914	5.2257	10	1.3817
0.1920	11°00′	.1908	.9816	.1944	5.1446	79°00′	1.3788
0.1949	10	.1937	.9811	.1974	5.0658	50	1.3759
0.1978	20	.1965	.9805	.2004	4.9894	40	1.3730
0.2007	30	.1994	.9799	.2035	4.9152	30	1.3701
0.2036	40	.2022	.9793	.2065	4.8430	20	1.3672
0.2065	50	.2051	.9787	.2095	4.7729	10	1.3643
0.2094	12°00′	.2079	.9781	.2126	4.7046	78°00′	1.3614
0.2123	10	.2108	.9775	.2156	4.6382	50	1.3584
0.2153	20	.2136	.9769	.2186	4.5736	40	1.3555
0.2182	30	.2164	.9763	.2217	4.5107	30	1.3526
0.2211	40	.2193	.9757	.2247	4.4494	20	1.3497
0.2240	50	.2221	.9750	.2278	4.3897	10	1.3468
0.2269	13°00′	.2250	.9744	.2309	4.3315	77°00′	1.3439
0.2298	10	.2278	.9737	.2339	4.2747	50	1.3410
0.2327	20	.2306	.9730	.2370	4.2193	40	1.3381
0.2356	30	.2334	.9724	.2401	4.1653	30	1.3352
0.2385	40	.2363	.9717	.2432	4.1126	20	1.3323
0.2414	50	.2391	.9710	.2462	4.0611	10	1.3294
0.2443	14°00′	.2419	.9703	.2493	4.0108	76°00′	1.3265
0.2473	10	.2447	.9696	.2524	3.9617	50	1.3235
0.2502	20	.2476	.9689	.2555	3.9136	40	1.3206
0.2531	30	.2504	.9681	.2586	3.8667	30	1.3177
0.2560	40	.2532	.9674	.2617	3.8208	20	1.3148
0.2589	50	.2560	.9667	.2648	3.7760	10	1.3119
0.2618	15°00′	.2588	.9659	.2679	3.7321	75°00′	1.3090
0.2647	10	.2616	.9652	.2711	3.6891	50	1.3061
0.2676	20	.2644	.9644	.2742	3.6470	40	1.3032
0.2705	30	.2672	.9636	.2773	3.6059	30	1.3003
0.2734	40	.2700	.9628	.2805	3.5656	20	1.2974
0.2763	50	.2728	.9621	.2836	3.5261	10	1.2945
0.2793	16°00′	.2756	.9613	.2867	3.4874	74°00′	1.2915
0.2822	10	.2784	.9605	.2899	3.4495	50	1.2886
0.2851	20	.2812	.9596	.2931	3.4124	40	1.2857
0.2880	30	.2840	.9588	.2962	3.3759	30	1.2828
0.2909	40	.2868	.9580	.2994	3.3402	20	1.2799
0.2938	50	.2896	.9572	.3026	3.3052	10	1.2770
0.2967	17°00′	.2924	.9563	.3057	3.2709	73°00′	1.2741
0.2996	10	.2952	.9555	.3089	3.2371	50	1.2712
0.3025	20	.2979	.9546	.3121	3.2041	40	1.2683
0.3054	30	.3007	.9537	.3153	3.1716	30	1.2654
0.3083	40	.3035	.9528	.3185	3.1397	20	1.2625
0.3113	50	.3062	.9520	.3217	3.1084	10	1.2595
0.3142	18°00′	.3090	.9511	.3249	3.0777	72°00′	1.2566
		Cosine	Sine	Cotangent	Tangent	Degrees	Radians

Table 2. (CONT.)

Radians	Degrees	Sine	Cosine	Tangent	Cotangent		
0.3142	18°00′	.3090	.9511	.3249	3.0777	72°00′	1.2566
0.3171	10	.3118	.9502	.3281	3.0475	50	1.2537
0.3200	20	.3145	.9492	.3314	3.0178	40	1.2508
0.3229	30	.3143	.9483	.3346	2.9887	30	1.2479
0.3258	40	.3201	.9474	.3378	2.9600	20	1.2450
0.3287	50	.3228	.9465	.3411	2.9319	10	1.2421
0.3316	19°00′	.3256	.9455	.3443	2.9042	71°00′	1.2392
0.3345	10	.3283	.9446	.3476	2.8770	50	1.2363
0.3374	20	.3311	.9436	.3508	2.8502	40	1.2334
0.3403	30	.3338	.9426	.3541	2.8239	30	1.2305
0.3432	40	.3365	.9417	.3574	2.7980	20	1.2275
0.3462	50	.3393	.9407	.3607	2.7725	10	1.2246
0.3491	20°00′	.3420	.9397	.3640	2.7475	70°00′	1.2217
0.3520	10	.3448	.9387	.3673	2.7228	50	1.2188
0.3549	20	.3475	.9377	.3706	2.6985	40	1.2159
0.3578	30	.3502	.9367	.3739	2.6746	30	1.2130
0.3607	40	.3529	.9356	.3772	2.6511	20	1.2101
0.3636	50	.3557	.9346	.3805	2.6279	10	1.2072
0.3655	21°00′	.3584	.9336	.3839	2.6051	69°00′	1.2043
0.3694	10	.3611	.9325	.3872	2.5826	50	1.2014
0.3723	20	.3638	.9315	.3906	2.5605	40	1.1985
0.3752	30	.3665	.9304	.3939	2.5386	30	1.1956
0.3782	40	.3692	.9293	.3973	2.5172	20	1.1926
0.3811	50	.3719	.9283	.4006	2.4960	10	1.1897
0.3840	22°00′	.3746	.9272	.4040	2.4751	68°00′	1.1868
0.3869	10	.3773	.9261	.4074	2.4545	50	1.1839
0.3898	20	.3800	.9250	.4108	2.4342	40	1.1810
0.3927	30	.3827	.9239	.4142	2.4142	30	1.1781
0.3956	40	.3854	.9228	.4176	2.3945	20	1.1752
0.3985	50	.3881	.9216	.4210	2.3750	10	1.1723
0.4014	23°00′	.3907	.9205	.4245	2.3559	67°00′	1.1694
0.4043	10	.3934	.9194	.4279	2.3369	50	1.1665
0.4072	20	.3961	.9182	.4314	2.3183	40	1.1636
0.4102	30	.3987	.9171	.4348	2.2998	30	1.1606
0.4131	40	.4014	.9159	.4383	2.2817	20	1.1577
0.4160	50	.4041	.9147	.4417	2.2637	10	1.1548
0.4189	24°00′	.4067	.9135	.4452	2.2460	66°00′	1.1519
0.4218	10	.4094	.9124	.4487	2.2286	50	1.1490
0.4247	20	.4120	.9112	.4522	2.2113	40	1.1461
0.4276	30	.4147	.9100	.4557	2.1943	30	1.1432
0.4305	40	.4173	.9088	.4592	2.1775	20	1.1403
0.4334	50	.4200	.9075	.4628	2.1609	10	1.1374
0.4363	25°00′	.4226	.9063	.4663	2.1445	65°00′	1.1345
0.4392	10	.4253	.9051	.4699	2.1283	50	1.1316
0.4422	20	.4279	.9038	.4734	2.1123	40	1.1286
0.4451	30	.4305	.9026	.4770	2.0965	30	1.1257
0.4480	40	.4331	.9013	.4806	2.0809	20	1.1228
0.4509	50	.4358	.9001	.4841	2.0655	10	1.1199
0.4538	26°00′	.4384	.8988	.4877	2.0503	64°00′	1.1170
0.4567	10	.4410	.8975	.4913	2.0353	50	1.1141
0.4596	20	.4436	.8962	.4950	2.0204	40	1.1112
0.4625	30	.4462	.8949	.4986	2.0057	30	1.1083
0.4654	40	.4488	.8936	.5022	1.9912	20	1.1054
0.4683	50	.4514	.8923	.5059	1.9768	10	1.1025
0.4712	27°00′	.4540	.8910	.5095	1.9626	63°00′	1.0996
		Cosine	Sine	Cotangent	Tangent	Degrees	Radians

Table 2. (Cont.)

Radians	Degrees	Sine	Cosine	Tangent	Cotangent		
0.4712	27°00′	.4540	.8910	.5095	1.9626	63°00′	1.0996
0.4741	10	.4566	.8897	.5132	1.9486	50	1.0966
0.4771	20	.4592	.8884	.5169	1.9347	40	1.0937
0.4800	30	.4617	.8870	.5206	1.9210	30	1.0908
0.4829	40	.4643	.8857	.5243	1.9074	20	1.0879
0.4858	50	.4669	.8843	.5280	1.8940	10	1.0850
0.4887	28°00′	.4695	.8829	.5317	1.8807	62°00′	1.0821
0.4916	10	.4720	.8816	.5354	1.8676	50	1.0792
0.4945	20	.4746	.8802	.5392	1.8546	40	1.0763
0.4974	30	.4772	.8788	.5430	1.8418	30	1.0734
0.5003	40	.4797	.8774	.5467	1.8291	20	1.0705
0.5032	50	.4823	.8760	.5505	1.8165	10	1.0676
0.5061	29°00′	.4848	.8746	.5543	1.8040	61°00′	1.0647
0.5091	10	.4874	.8732	.5581	1.7917	50	1.0617
0.5120	20	.4899	.8718	.5619	1.7796	40	1.0588
0.5149	30	.4924	.8704	.5658	1.7675	30	1.0559
0.5178	40	.4950	.8689	.5696	1.7556	20	1.0530
0.5207	50	.4975	.8675	.5735	1.7437	10	1.0501
0.5236	30°00′	.5000	.8660	.5774	1.7321	60°00′	1.0472
0.5265	10	.5025	.8646	.5812	1.7205	50	1.0443
0.5294	20	.5050	.8631	.5851	1.7090	40	1.0414
0.5323	30	.5075	.8616	.5890	1.6977	30	1.0385
0.5352	40	.5100	.8601	.5930	1.6864	20	1.0356
0.5381	50	.5125	.8587	.5969	1.6753	10	1.0327
0.5411	31°00′	.5150	.8572	.6009	1.6643	59°00′	1.0297
0.5440	10	.5175	.8557	.6048	1.6534	50	1.0268
0.5469	20	.5200	.8542	.6088	1.6426	40	1.0239
0.5498	30	.5225	.8526	.6128	1.6319	30	1.0210
0.5527	40	.5250	.8511	.6168	1.6212	20	1.0181
0.5556	50	.5275	.8496	.6208	1.6107	10	1.0152
0.5585	32°00′	.5299	.8480	.6249	1.6003	58°00′	1.0123
0.5614	10	.5324	.8465	.6289	1.5900	50	1.0094
0.5643	20	.5348	.8450	.6330	1.5798	40	1.0065
0.5672	30	.5373	.8434	.6371	1.5697	30	1.0036
0.5701	40	.5398	.8418	.6412	1.5597	20	1.0007
0.5730	50	.5422	.8403	.6453	1.5497	10	0.9977
0.5760	33°00′	.5446	.8387	.6494	1.5399	57°00′	0.9948
0.5789	10	.5471	.8371	.6536	1.5301	50	0.9919
0.5818	20	.5495	.8355	.6577	1.5204	40	0.9890
0.5847	30	.5519	.8339	.6619	1.5108	30	0.9861
0.5876	40	.5544	.8323	.6661	1.5013	20	0.9832
0.5905	50	.5568	.8307	.6703	1.4919	10	0.9803
0.5934	34°00′	.5592	.8290	.6745	1.4826	56°00′	0.9774
0.5963	10	.5616	.8274	.6787	1.4733	50	0.9745
0.5992	20	.5640	.8258	.6830	1.4641	40	0.9716
0.6021	30	.5664	.8241	.6873	1.4550	30	0.9687
0.6050	40	.5688	.8225	.6916	1.4460	20	0.9657
0.6080	50	.5712	.8208	.6959	1.4370	10	0.9628
0.6109	35°00′	.5736	.8192	.7002	1.4281	55°00′	0.9599
0.6138	10	.5760	.8175	.7046	1.4193	50	0.9570
0.6167	20	.5783	.8158	.7089	1.4106	40	0.9541
0.6196	30	.5807	.8141	.7133	1.4019	30	0.9512
0.6225	40	.5831	.8124	.7177	1.3934	20	0.9483
0.6254	50	.5854	.8107	.7221	1.3848	10	0.9454
0.6283	36°00′	.5878	.8090	.7265	1.3764	54°00′	0.9425
		Cosine	Sine	Cotangent	Tangent	Degrees	Radians

Table 2. (CONT.)

Radians	Degrees	Sine	Cosine	Tangent	Cotangent		
0.6283	36°00′	.5878	.8090	.7265	1.3764	54°00′	0.9425
0.6312	10	.5901	.8073	.7310	1.3680	50	0.9396
0.6341	20	.5925	.8056	.7355	1.3597	40	0.9367
0.6370	30	.5948	.8039	.7400	1.3514	30	0.9338
0.6400	40	.5972	.8021	.7445	1.3432	20	0.9308
0.6429	50	.5995	.8004	.7490	1.3351	10	0.9279
0.6458	37°00′	.6018	.7986	.7536	1.3270	53°00′	0.9250
0.6487	10	.6041	.7969	.7581	1.3190	50	0.9221
0.6516	20	.6065	.7951	.7627	1.3111	40	0.9192
0.6545	30	.6088	.7934	.7673	1.3032	30	0.9163
0.6574	40	.6111	.7916	.7720	1.2954	20	0.9134
0.6603	50	.6134	.7898	.7766	1.2876	10	0.9105
0.6632	38°00′	.6157	.7880	.7813	1.2799	52°00′	0.9076
0.6661	10	.6180	.7862	.7860	1.2723	50	0.9047
0.6690	20	.6202	.7844	.7907	1.2647	40	0.9018
0.6720	30	.6225	.7826	.7954	1.2572	30	0.8988
0.6749	40	.6248	.7808	.8002	1.2497	20	0.8959
0.6778	50	.6271	.7790	.8050	1.2423	10	0.8930
0.6807	39°00′	.6293	.7771	.8098	1.2349	51°00′	0.8901
0.6836	10	.6316	.7753	.8146	1.2276	50	0.8872
0.6865	20	.6338	.7735	.8195	1.2203	40	0.8843
0.6894	30	.6361	.7716	.8243	1.2131	30	0.8814
0.6923	40	.6383	.7698	.8292	1.2059	20	0.8785
0.6952	50	.6406	.7679	.8342	1.1988	10	0.8756
0.6981	40°00′	.6428	.7660	.8391	1.1918	50°00′	0.8727
0.7010	10	.6450	.7642	.8441	1.1847	50	0.8698
0.7039	20	.6472	.7623	.8491	1.1778	40	0.8668
0.7069	30	.6494	.7604	.8541	1.1708	30	0.8639
0.7098	40	.6517	.7585	.8591	1.1640	20	0.8610
0.7127	50	.6539	.7566	.8642	1.1571	10	0.8581
0.7156	41°00′	.6561	.7547	.8693	1.1504	49°00′	0.8552
0.7185	10	.6583	.7528	.8744	1.1436	50	0.8523
0.7214	20	.6604	.7509	.8796	1.1369	40	0.8494
0.7243	30	.6626	.7490	.8847	1.1303	30	0.8465
0.7272	40	.6648	.7470	.8899	1.1237	20	0.8436
0.7301	50	.6670	.7451	.8952	1.1171	10	0.8407
0.7330	42°00′	.6691	.7431	.9004	1.1106	48°00′	0.8378
0.7359	10	.6713	.7412	.9057	1.1041	50	0.8348
0.7389	20	.6734	.7392	.9110	1.0977	40	0.8319
0.7418	30	.6756	.7373	.9163	1.0913	30	0.8290
0.7447	40	.6777	.7353	.9217	1.0850	20	0.8261
0.7476	50	.6799	.7333	.9271	1.0786	10	0.8232
0.7505	43°00′	.6820	.7314	.9325	1.0724	47°00′	0.8203
0.7534	10	.6841	.7294	.9380	1.0661	50	0.8174
0.7563	20	.6862	.7274	.9435	1.0599	40	0.8145
0.7592	30	.6884	.7254	.9490	1.0538	30	0.8116
0.7621	40	.6905	.7234	.9545	1.0477	20	0.8087
0.7650	50	.6926	.7214	.9601	1.0416	10	0.8058
0.7679	44°00′	.6947	.7193	.9657	1.0355	46°00′	0.8029
0.7709	10	.6967	.7173	.9713	1.0295	50	0.7999
0.7738	20	.6988	.7153	.9770	1.0235	40	0.7970
0.7767	30	.7009	.7133	.9827	1.0176	30	0.7941
0.7796	40	.7030	.7112	.9884	1.0117	20	0.7912
0.7825	50	.7050	.7092	.9942	1.0058	10	0.7883
0.7854	45°00′	.7071	.7071	1.0000	1.0000	45°00′	0.7854
		Cosine	Sine	Cotangent	Tangent	Degrees	Radians

Table 3. NATURAL OR NAPIERIAN LOGS†

0.000–0.499

x	0	1	2	3	4	5	6	7	8	9
0.00	$-\infty$	−6 .90776	−6 .21461	−5 .80914	−5 .52146	−5 .29832	−5 .11600	−4 .96185	−4 .82831	−4 .71053
.01	−4.60517	.50986	.42285	.34281	.26870	.19971	.13517	.07454	.01738	*.96332
.02	−3.91202	.86323	.81671	.77226	.72970	.68888	.64966	.61192	.57555	.54046
.03	.50656	.47377	.44202	.41125	.38139	.35241	.32424	.29684	.27017	.24419
.04	.21888	.19418	.17009	.14656	.12357	.10109	.07911	.05761	.03655	.01593
.05	−2.99573	.97593	.95651	.93746	.91877	.90042	.88240	.86470	.84731	.83022
.06	.81341	.79688	.78062	.76462	.74887	.73337	.71810	.70306	.68825	.67365
.07	.65926	.64508	.63109	.61730	.60369	.59027	.57702	.56395	.55105	.53831
.08	.52573	.51331	.50104	.48891	.47694	.46510	.45341	.44185	.43042	.41912
.09	.40795	.39690	.38597	.37516	.36446	.35388	.34341	.33304	.32279	.31264
0.10	−2.30259	.29263	.28278	.27303	.26336	.25379	.24432	.23493	.22562	.21641
.11	.20727	.19823	.18926	.18037	.17156	.16282	.15417	.14558	.13707	.12863
.12	.12026	.11196	.10373	.09557	.08747	.07944	.07147	.06357	.05573	.04794
.13	.04022	.03256	.02495	.01741	.00992	.00248	*.99510	*.98777	*.98050	*.97328
.14	−1.96611	.95900	.95193	.94491	.93794	.93102	.92415	.91732	.91054	.90381
.15	.89712	.89048	.88387	.87732	.87080	.86433	.85790	.85151	.84516	.83885
.16	.83258	.82635	.82016	.81401	.80789	.80181	.79577	.78976	.78379	.77786
.17	.77196	.76609	.76026	.75446	.74870	.74297	.73727	.73161	.72597	.72037
.18	.71480	.70926	.70375	.69827	.69282	.68740	.68201	.67665	.67131	.66601
.19	.66073	.65548	.65026	.64507	.63990	.63476	.62964	.62455	.61949	.61445
0.20	−1.60944	.60445	.59949	.59455	.58964	.58475	.57988	.57504	.57022	.56542
.21	.56065	.55590	.55117	.54646	.54178	.53712	.53248	.52786	.52326	.51868
.22	.51413	.50959	.50508	.50058	.49611	.49165	.48722	.48281	.47841	.47403
.23	.46968	.46534	.46102	.45672	.45243	.44817	.44392	.43970	.43548	.43129
.24	.42712	.42296	.41882	.41469	.41059	.40650	.40242	.39837	.39433	.39030
.25	.38629	.38230	.37833	.37437	.37042	.36649	.36258	.35868	.35480	.35093
.26	.34707	.34323	.33941	.33560	.33181	.32803	.32426	.32051	.31677	.31304
.27	.30933	.30564	.30195	.29828	.29463	.29098	.28735	.28374	.28013	.27654
.28	.27297	.26940	.26585	.26231	.25878	.25527	.25176	.24827	.24479	.24133
.29	.23787	.23443	.23100	.22758	.22418	.22078	.21740	.21402	.21066	.20731
0.30	−1.20397	.20065	.19733	.19402	.19073	.18744	.18417	.18091	.17766	.17441
.31	.17118	.16796	.16475	.16155	.15836	.15518	.15201	.14885	.14570	.14256
.32	.13943	.13631	.13320	.13010	.12701	.12393	.12086	.11780	.11474	.11170
.33	.10866	.10564	.10262	.09961	.09661	.09362	.09064	.08767	.08471	.08176
.34	.07881	.07587	.07294	.07002	.06711	.06421	.06132	.05843	.05555	.05268
.35	−1.04982	.04697	.04412	.04129	.03846	.03564	.03282	.03002	.02722	.02443
.36	.02165	.01888	.01611	.01335	.01060	.00786	.00512	.00239	*.99967	*.99696
.37	−0.99425	.99155	.98886	.98618	.98350	.98083	.97817	.97551	.97286	.97022
.38	.96758	.96496	.96233	.95972	.95711	.95451	.95192	.94933	.94675	.94418
.39	.94161	.93905	.93649	.93395	.93140	.92887	.92634	.92382	.92130	.91879
0.40	−0.91629	.91379	.91130	.90882	.90634	.90387	.90140	.89894	.89649	.89404
.41	.89160	.88916	.88673	.88431	.88189	.87948	.87707	.87467	.87227	.86988
.42	.86750	.86512	.86275	.86038	.85802	.85567	.85332	.85097	.84863	.84630
.43	.84397	.84165	.83933	.83702	.83471	.83241	.83011	.82782	.82554	.82326
.44	.82098	.81871	.81645	.81419	.81193	.80968	.80744	.80520	.80296	.80073
.45	.79851	.79629	.79407	.79186	.78966	.78746	.78526	.78307	.78089	.77871
.46	.77653	.77436	.77219	.77003	.76787	.76572	.76357	.76143	.75929	.75715
.47	.75502	.75290	.75078	.74866	.74655	.74444	.74234	.74024	.73814	.73605
.48	.73397	.73189	.72981	.72774	.72567	.72361	.72155	.71949	.71744	.71539
.49	.71335	.71131	.70928	.70725	.70522	.70320	.70118	.69917	.69716	.69515

†All numbers in this table are negative.

Table 3. (CONT.)†

0.500–1.000

x	0	1	2	3	4	5	6	7	8	9
0.50	−0.69315	.69115	.68916	.68717	.68518	.68320	.68122	.67924	.67727	.67531
.51	.67334	.67139	.66943	.66748	.66553	.66359	.66165	.65971	.65778	.65585
.52	.65393	.65201	.65009	.64817	.64626	.64436	.64245	.64055	.63866	.63677
.53	.63488	.63299	.63111	.62923	.62736	.62549	.62362	.62176	.61990	.61804
.54	.61619	.61434	.61249	.61065	.60881	.60697	.60514	.60331	.60148	.59966
.55	.59784	.59602	.59421	.59240	.59059	.58879	.58699	.58519	.58340	.58161
.56	.57982	.57803	.57625	.57448	.57270	.57093	.56916	.56740	.56563	.56387
.57	.56212	.56037	.55862	.55687	.55513	.55339	.55165	.54991	.54818	.54645
.58	.54473	.54300	.54128	.53957	.53785	.53614	.53444	.53273	.53103	.52933
.59	.52763	.52594	.52425	.52256	.52088	.51919	.51751	.51584	.51416	.51249
0.60	−0.51083	.50916	.50750	.50584	.50418	.50253	.50088	.49923	.49758	.49594
.61	.49430	.49266	.49102	.48939	.48776	.48613	.48451	.48289	.48127	.47965
.62	.47804	.47642	.47482	.47321	.47160	.47000	.46840	.46681	.46522	.46362
.63	.46204	.46045	.45887	.45728	.45571	.45413	.45256	.45099	.44942	.44785
.64	.44629	.44473	.44317	.44161	.44006	.43850	.43696	.43541	.43386	.43232
.65	.43078	.42925	.42771	.42618	.42465	.42312	.42159	.42007	.41855	.41703
.66	.41552	.41400	.41249	.41098	.40947	.40797	.40647	.40497	.40347	.40197
.67	.40048	.39899	.39750	.39601	.39453	.39304	.39156	.39008	.38861	.38713
.68	.38566	.38419	.38273	.38126	.37980	.37834	.37688	.37542	.37397	.37251
.69	.37106	.36962	.36817	.36673	.36528	.36384	.36241	.36097	.35954	.35810
0.70	−0.35667	.35525	.35382	.35240	.35098	.34956	.34814	.34672	.34531	.34390
.71	.34249	.34108	.33968	.33827	.33687	.33547	.33408	.33268	.33129	.32989
.72	.32850	.32712	.32573	.32435	.32296	.32158	.32021	.31883	.31745	.31608
.73	.31471	.31334	.31197	.31061	.30925	.30788	.30653	.30517	.30381	.30246
.74	.30111	.29975	.29841	.29706	.29571	.29437	.29303	.29169	.29035	.28902
.75	.28768	.28635	.28502	.28369	.28236	.28104	.27971	.27839	.27707	.27575
.76	.27444	.27312	.27181	.27050	.26919	.26788	.26657	.26527	.26397	.26266
.77	.26136	.26007	.25877	.25748	.25618	.25489	.25360	.25231	.25103	.24974
.78	.24846	.24718	.24590	.24462	.24335	.24207	.24080	.23953	.23826	.23699
.79	.23572	.23446	.23319	.23193	.23067	.22941	.22816	.22690	.22565	.22439
0.80	−0.22314	.22189	.22065	.21940	.21816	.21691	.21567	.21443	.21319	.21196
.81	.21072	.20949	.20825	.20702	.20579	.20457	.20334	.20212	.20089	.19967
.82	.19845	.19723	.19601	.19480	.19358	.19237	.19116	.18995	.18874	.18754
.83	.18633	.18513	.18392	.18272	.18152	.18032	.17913	.17793	.17674	.17554
.84	.17435	.17316	.17198	.17079	.16960	.16842	.16724	.16605	.16487	.16370
.85	−0.16252	.16134	.16017	.15900	.15782	.15665	.15548	.15432	.15315	.15199
.86	.15082	.14966	.14850	.14734	.14618	.14503	.14387	.14272	.14156	.14041
.87	.13926	.13811	.13697	.13582	.13467	.13353	.13239	.13125	.13011	.12897
.88	.12783	.12670	.12556	.12443	.12330	.12217	.12104	.11991	.11878	.11766
.89	.11653	.11541	.11429	.11317	.11205	.11093	.10981	.10870	.10759	.10647
0.90	−0.10536	.10425	.10314	.10203	.10093	.09982	.09872	.09761	.09651	.09541
.91	.09431	.09321	.09212	.09102	.08992	.08883	.08774	.08665	.08556	.08447
.92	.08338	.08230	.08121	.08013	.07904	.07796	.07688	.07580	.07472	.07365
.93	.07257	.07150	.07042	.06935	.06828	.06721	.06614	.06507	.06401	.06294
.94	.06188	.06081	.05975	.05869	.05763	.05657	.05551	.05446	.05340	.05235
.95	.05129	.05024	.04919	.04814	.04709	.04604	.04500	.04395	.04291	.04186
.96	.04082	.03978	.03874	.03770	.03666	.03563	.03459	.03356	.03252	.03149
.97	.03046	.02943	.02840	.02737	.02634	.02532	.02429	.02327	.02225	.02122
.98	.02020	.01918	.01816	.01715	.01613	.01511	.01410	.01309	.01207	.01106
.99	.01005	.00904	.00803	.00702	.00602	.00501	.00401	.00300	.00200	.00100
1.00	0.00000									

†All numbers in this table are negative.

Table 4. Exponentials e^x and e^{-x}

x	e^x	x	e^x	x	e^x	x	e^{-x}	x	e^{-x}	x	e^{-x}
0.00	1.000	**0.50**	1.649	**1.0**	2.718	**0.00**	1.000	**0.50**	.607	**1.0**	.368
.01	1.010	.51	1.665	.1	3.004	.01	0.990	.51	.600	.1	.333
.02	1.020	.52	1.682	.2	3.320	.02	.980	.52	.595	.2	.301
.03	1.030	.53	1.699	.3	3.669	.03	.970	.53	.589	.3	.273
.04	1.041	.54	1.716	.4	4.055	.04	.961	.54	.583	.4	.247
0.05	1.051	0.55	1.733	1.5	4.482	0.05	.951	0.55	.577	1.5	.223
.06	1.062	.56	1.751	.6	4.953	.06	.942	.56	.571	.6	.202
.07	1.073	.57	1.768	.7	5.474	.07	.932	.57	.566	.7	.183
.08	1.083	.58	1.786	.8	6.050	.08	.923	.58	.560	.8	.165
.09	1.094	.59	1.804	.9	6.686	.09	.914	.59	.554	.9	.150
0.10	1.105	**0.60**	1.822	**2.0**	7.389	**0.10**	.905	**0.60**	.549	**2.0**	.135
.11	1.116	.61	1.840	.1	8.166	.11	.896	.61	.543	.1	.122
.12	1.127	.62	1.859	.2	9.025	.12	.887	.62	.538	.2	.111
.13	1.139	.63	1.878	.3	9.974	.13	.878	.63	.533	.3	.100
.14	1.150	.64	1.896	.4	11.02	.14	.869	.64	.527	.4	.0907
0.15	1.162	0.65	1.916	2.5	12.18	0.15	.861	0.65	.522	2.5	.0821
.16	1.174	.66	1.935	.6	13.46	.16	.852	.66	.517	.6	.0743
.17	1.185	.67	1.954	.7	14.88	.17	.844	.67	.512	.7	.0672
.18	1.197	.68	1.974	.8	16.44	.18	.835	.68	.507	.8	.0608
.19	1.209	.69	1.994	.9	18.17	.19	.827	.69	.502	.9	.0550
0.20	1.221	**0.70**	2.014	**3.0**	20.09	**0.20**	.819	**0.70**	.497	**3.0**	.0498
.21	1.234	.71	2.034	.1	22.20	.21	.811	.71	.492	.1	.0450
.22	1.246	.72	2.054	.2	24.53	.22	.803	.72	.487	.2	.0408
.23	1.259	.73	2.075	.3	27.11	.23	.795	.73	.482	.3	.0369
.24	1.271	.74	2.096	.4	29.96	.24	.787	.74	.477	.4	.0334
0.25	1.284	0.75	2.117	3.5	33.12	0.25	.779	0.75	.472	3.5	.0302
.26	1.297	.76	2.138	.6	36.60	.26	.771	.76	.468	.6	.0273
.27	1.310	.77	2.160	.7	40.45	.27	.763	.77	.463	.7	.0247
.28	1.323	.78	2.181	.8	44.70	.28	.756	.78	.458	.8	.0224
.29	1.336	.79	2.203	.9	49.40	.29	.748	.79	.454	.9	.0202
0.30	1.350	**0.80**	2.226	**4.0**	54.60	**0.30**	.741	**0.80**	.449	**4.0**	.0183
.31	1.363	.81	2.248	.1	60.34	.31	.733	.81	.445	.1	0.166
.32	1.377	.82	2.270	.2	66.69	.32	.726	.82	.440	.2	.0150
.33	1.391	.83	2.293	.3	73.70	.33	.719	.83	.436	.3	.0136
.34	1.405	.84	2.316	.4	81.45	.34	.712	.84	.432	.4	.0123
0.35	1.419	0.85	2.340	4.5	90.02	0.35	.705	0.85	.427	4.5	.0111
.36	1.433	.86	2.363			.36	.698	.86	.423		
.37	1.448	.87	2.387	**5.0**	148.4	.37	.691	.87	.419	**5.0**	.00674
.38	1.462	.88	2.411	6.0	403.4	.38	.684	.88	.415	6.0	.00248
.39	1.477	.89	2.435	7.0	1097.	.39	.677	.89	.411	7.0	.000912
0.40	1.492	**0.90**	2.460	8.0	2981.	**0.40**	.670	**0.90**	.407	8.0	.000335
.41	1.507	.91	2.484	9.0	8103.	.41	.664	.91	.403	9.0	.000123
.42	1.522	.92	2.509	**10.0**	22026.	.42	.657	.92	.399	**10.0**	.000045
.43	1.537	.93	2.535			.43	.651	.93	.395		
.44	1.553	.94	2.560			.44	.644	.94	.391		
0.45	1.568	0.95	2.586			0.45	.638	0.95	.387		
.46	1.584	.96	2.612			.46	.631	.96	.383		
.47	1.600	.97	2.638			.47	.625	.97	.379		
.48	1.616	.98	2.664			.48	.619	.98	.375		
.49	1.632	.99	2.691			.49	.613	.99	.372		
0.50	1.649	**1.00**	2.718			**0.50**	0.607	**1.00**	.368		

INDEX